IMAGES
of America

MYRTLE CREEK

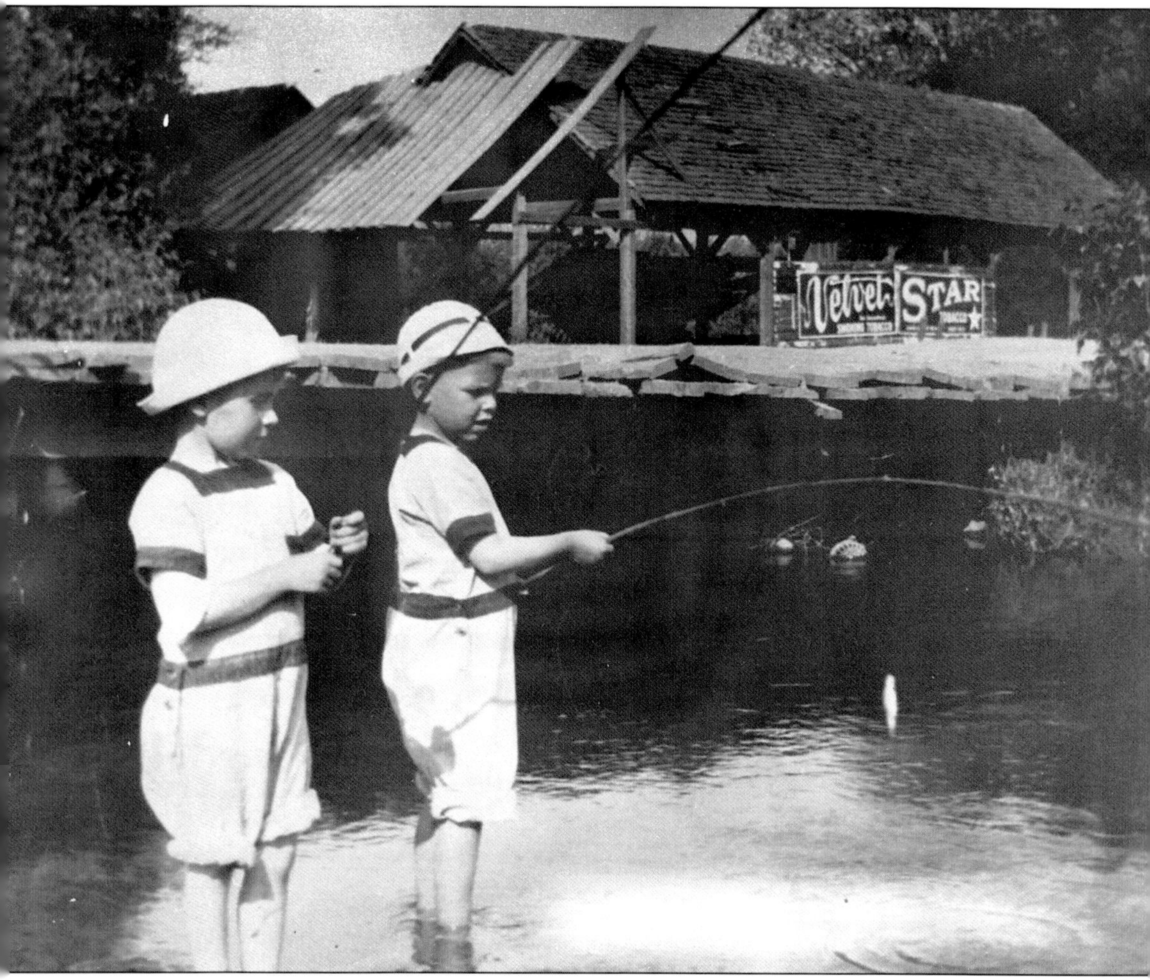

Around 1915, cousins Vinton Hall (left) and Hall Seely, grandsons of John and Susanna Hall, fish in Myrtle Creek near Days Creek Cutoff Road. Hall Seely, the son of Dr. A.C. and Florence Hall Seely, was born in 1907 and became a nationally prominent surgeon with a practice in Roseburg. Related to the Halls, Riddles, and Weavers, the Seelys spent many weekends in Myrtle Creek. (Courtesy of Douglas County Museum.)

On the Cover: This crew is busy at work on farmland near Neal Lane. The flume behind them was built in 1903 for $25,000 with 2.5 million board feet of lumber and reached 50 feet in height. It ran six miles up South Myrtle to the Johnson mill and was used to float lumber to a planer mill and sorting shed. The mill was located on a railroad spur on the south side of present-day Millsite Park. (Courtesy of City of Myrtle Creek.)

IMAGES
of America

MYRTLE CREEK

Myrtle Creek Historical Society
Foreword by Mayor Dan Jocoy

ARCADIA
PUBLISHING

ISBN 978-0-7385-9577-1

Published by Arcadia Publishing
Charleston, South Carolina

Printed in the United States of America

Library of Congress Control Number: 2012935608

For all general information, please contact Arcadia Publishing:
Telephone 843-853-2070
Fax 843-853-0044
E-mail sales@arcadiapublishing.com
For customer service and orders:
Toll-Free 1-888-313-2665

Visit us on the Internet at www.arcadiapublishing.com

Dedicated to Maureen Butler, who was the driving force behind the creation of this book; to Dale Greenley, the tireless researcher; and to all our esteemed pioneers.

Contents

FOREWORD

Would you take a quick journey with me? Let's take an imaginary walk and visit some of the people and places of our town called Myrtle Creek. We'll stroll up Main Street, where first we meet John and Susanna Hall. The Halls are considered the founders of Myrtle Creek. Farther up Main was the Dyer home. Moses True Dyer arrived here in 1852.

Turn up Third Avenue and sit on the porch with Susan Dailey. She traveled as a child over the Oregon Trail with her parents, Dr. Benjamin Fallin and his wife, Sarah. It is said that Susan made house calls with her father on horse and buggy.

The next stop on Third Avenue is the Selig home, which is now our home. When we purchased this historic house, my wife and I felt a little of that same excitement pioneers had traveling west in the Great Migration of 1843. We often wish the Selig family could come and see how we restored their home into a bed and breakfast and tearoom. Each room is named after one of their family members. Among the photographs in this book, you will see the children as they pose in front of their school, which was across the street on Pleasant Street.

We have enthusiastically researched the history of the Selig home and, because I am both the town's mayor and a local pastor, I desired to know the names and stories of the people who established Myrtle Creek. I wanted to know how they lived, their passions in life, and how they died.

Yet, something more interesting occurred along the way. I formed deeper relationships with long-standing community members. What I discovered were people who sacrificed for the good of the community. Many were brave and stood tall during very difficult times. I heard of good people who were community-minded and hardworking.

Thanks to the dedication of the Myrtle Creek Historical Society, you have at your fingertips a glimpse into the lives of generations of this small, southern Oregon logging and farming town we call home. Let's turn the pages of this photographic record and journey together. Let's explore what it was like to live during a much harder, but much simpler, time.

Dan Jocoy
Mayor, City of Myrtle Creek

ACKNOWLEDGMENTS

The Myrtle Creek Historical Society has many people and organizations to thank for their help with the assembly of the photographs and information contained in these pages. Initially, we would like to thank Maureen Butler for taking the lead in this endeavor; her leadership was invaluable. Additional thanks to Dale Greenley, who spent countless hours and many months in the museum basement searching the archives. Plus a genuine thank-you to everyone who graciously shared their photographs for this project.

Photographs donated by private citizens are credited in the caption. All other photographs are courtesy of the Douglas County Museum of History and Natural History. The museum staff, in particular photo archivist Jena Mitchell and research librarian Karen Bratton, were more than generous, giving cheerful help and providing the resources necessary to compile the photographs and historic information used in this book.

Also providing assistance and information were the City of Myrtle Creek officials and office staff, Steve Gorthy and the Cascade Historical Motor Club of Roseburg, and the Pioneer Museum in Canyonville, notably member Lillian Stevenson. A special thank-you goes to Dave Lines of Canyonville. He provided photographic expertise and the use of his computer and scanner to scan the hundreds of photographs brought to our meetings by generous citizens.

This book is primarily a book of photographs, but the historical society wanted to include as much local history as we could weave into the captions. Much of the oral history was provided by Blanche Newton and her invaluable book, *Golden Leafs of Time*, as well as the recollections of Bill Rice, Dora Dyer, Al Helgeson, Nancy Johnson, Darrell Bangs, Becky Norton, and Bill Leming. We extend our thanks to them and the many people who attended our meetings, donated the use of their personal photographs, and related many interesting tales of yesteryear. We wish there was room in these pages to include every memory shared, but it just was not possible. Tears were shed as we realized many wonderful photographs must be excluded and that many old photographs were lost to time, never to be seen again.

Thank you for letting us take this opportunity to share the story of our town with you.

Myrtle Creek Historical Society
www.MyrtleCreekHistory.com

Introduction

Myrtle Creek is located in southern Oregon along Interstate 5. It was named after the creek that flows through the valley and the Myrtle trees that grow abundantly in the area. In 1846, a survey party led by Jesse Applegate and Levi Scott searched for an easier and less hazardous route for pioneers headed to Oregon. The route they scouted became known as the Scott-Applegate Trail, and part of it followed the South Umpqua River through the future site of Myrtle Creek. The longest undisturbed remnant of the trail remaining in Oregon can still be seen near town at the top of the hill on Dole Road.

Myrtle Creek is one of the earliest settlements in southern Oregon. It is the site of the first gristmill in southern Oregon and the first multi-arch bridge designed by noted bridge engineer Conde B. McCullough. Another bridge of note is the Neal Lane covered bridge across South Myrtle Creek, reputed to be the shortest covered bridge in the state. Flowing beside the town is the South Umpqua River, known for gold mining, fishing, and beautiful scenery. Paul Jenkins wrote of Myrtle Creek in a 1935 Roseburg newspaper:

> Located in one of the richest sections of Douglas County, the town of Myrtle Creek is a thriving business center, drawing its trade from a wide area. To the south of it lies the Missouri Bottom country, a valley four or five miles long, through which runs the South Umpqua River. To the northeast and east are the two valleys of North Myrtle and South Myrtle creeks, each ten or twelve miles in length . . . narrow and winding, but very fertile.

Following the California gold rush and the passage of the Donation Land Claim Act in 1850, a wave of early pioneers brought William Weaver, John W. Weaver, James Weaver, Henry A. Adams, and their families from Missouri to settle in this valley, which they called Missouri Bottom. James Weaver, who had filed his claim on the future town site, sold it in 1851 to J. Bailey for a yoke of oxen. Bailey, in turn, sold the claim to Lazarus Wright, who built the first gristmill in this part of the state. The millstones were sailed around Cape Horn to Scottsburg and freighted to Myrtle Creek. For years, the mill supplied settlers with their grain products. In August 1930, the "old mill" was torn down to make room for the Pacific Highway Bridge over Myrtle Creek. A millstone has survived and is now on display at Millsite Park.

Wright also operated a hotel and was the first postmaster when the Myrtle Creek Post Office was officially dedicated on February 18, 1854. In 1865, Wright sold 320 of his acres, including the gristmill, to John Hall. In 1868, Hall had the town site platted and subdivided. Although he was not one of the original settlers, Hall is considered the founder of Myrtle Creek.

Gold was discovered in the South Umpqua Valley, and prospectors and settlers rushed to Myrtle Creek's rich and fertile valleys. Over $100,000 worth of gold was extracted from the Continental Mine alone. Stores sprang up along the main streets of the town site to supply the miners and settlers. Shops included Hunsaker Mercantile; Marks, Wollenburg and Company; Adams and Company;

and H. Dyer and F.M. Gabbert's Farmers Mercantile Establishment. The hills and tributaries of the South Umpqua River still yield enough gold to keep amateur gold seekers happy.

In 1859, the stagecoach came to Myrtle Creek. Shortly afterwards, the telegraph came to town, and in 1882, the railroad arrived. In 1889, the longest hydrologic placer mining system in Douglas County was conceived. The 27-mile-long project employed many Chinese laborers and is often referred to as the "China Ditch." It excited the interest of local residents, and many invested money, but it was never finished.

Myrtle Creek incorporated as a city in 1893. In 1902, Thurman Chaney and Charles Rice started a newspaper, the *Myrtle Creek Mail*. It has been published continuously except for a short hiatus during World War II. Relatives of Thurman Chaney publish the town's local newspaper today.

A time of prosperity followed the turn of the 20th century. By 1910, eight more subdivisions had been added to the city. New public schools were built, and several new churches were established. Gold and lumber were both significant features of the economy. Dozens of lumber mills dotted the landscape up the North and South Myrtle Creek Valleys, and timber was as important then as it is today.

For nearly 40 years—roughly from 1895 to 1935—prunes were the chief agricultural product of Douglas County, and Myrtle Creek was the county's leading producer. John Hall, Jake Chadwick, John Weaver, Henry Adams, Hans Weaver, and Ed Weaver were among the important Douglas County prune growers. By the mid-1930s, the once-booming prune industry had faded, leaving Myrtle Creek to suffer through the Depression until the postwar building boom began.

In 1922, Pacific Highway was completed and the north entrance to town was relocated via an extension of Main Street to the new South Umpqua Bridge. Oregon's master bridge engineer, Conde McCullough, designed his first multi-arch bridge to serve as the elegant new entry into Myrtle Creek. This major north-south route brought tourist dollars into town, and service stations emerged on nearly every block as motorcars became the modern mode of transportation.

After the war, the rapidly growing timber industry and a population explosion led to a local housing shortage. A tent city housed many homeless families until 25 new subdivisions were added to the city, eight of these in 1947 alone. For entertainment, the community boasted a movie theater, a bowling alley, soda fountains, a drive-in theater, Chap's Place, a Dairy Queen, and more, all thriving in the booming economy.

Despite some major floods, which changed the face of downtown, the city continued to grow and prosper. During the 1950s, a new city hall was built, the police and fire departments were expanded, a public pool and library were opened, and a new hospital began serving the community.

Due to growing interest in our history, the Myrtle Creek Historical Society was formed. This book reflects the efforts of the historical society to gather the scattered photographs and documents related to the City of Myrtle Creek and adjoining unincorporated communities of Tri-City, North and South Myrtle, Ruckles, Dole, and others. It is our mission to preserve local history for future generations. Another group of volunteers is dedicated to restoration of the Pioneer Cemetery.

We hope you will enjoy the tour as we attempt to take you back through the first 100 years of settlement and growth, memories of which are but faint shadows of a time since passed.

Natural-born curiosity should inspire one to look from whence we came in order to better prepare for where we are going.

—Martin Sheen

One

For a Yoke of Oxen, a City is Born

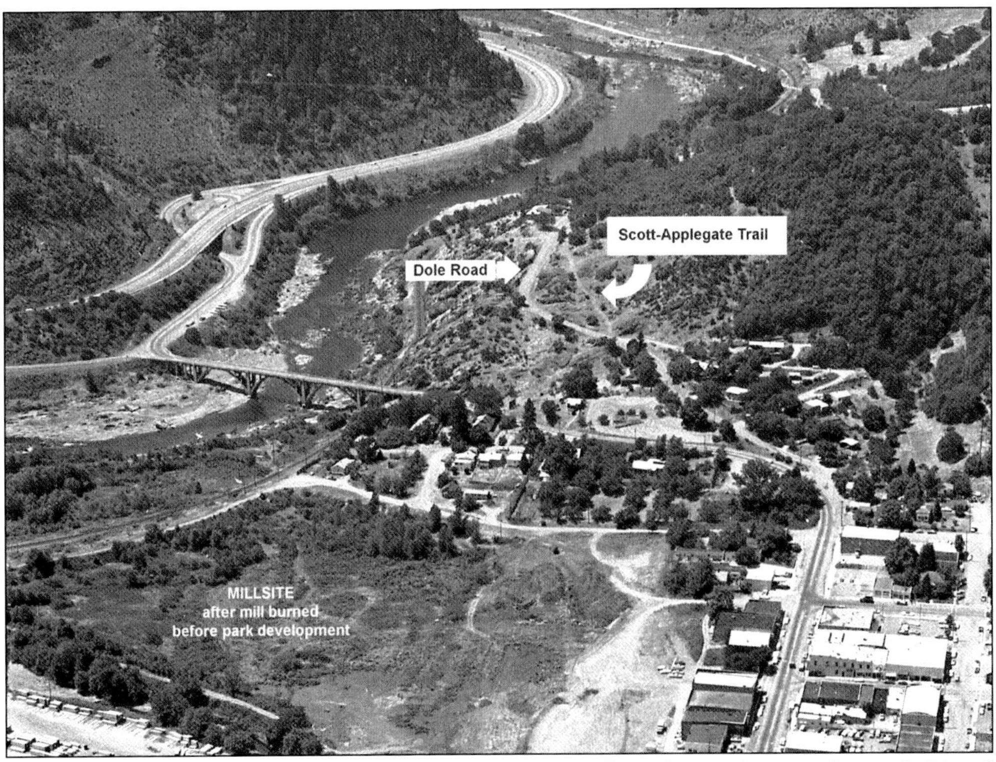

In 1846, Levi Scott led the first wagon train into Oregon from the south using the newly blazed Scott-Applegate Trail. Until the railroad came in 1882, this trail, improved as a military road in the 1850s, served as the only route into Myrtle Creek. In this 1973 aerial photograph, an arrow points to the longest undisturbed section of the Scott-Applegate Trail remaining in Oregon.

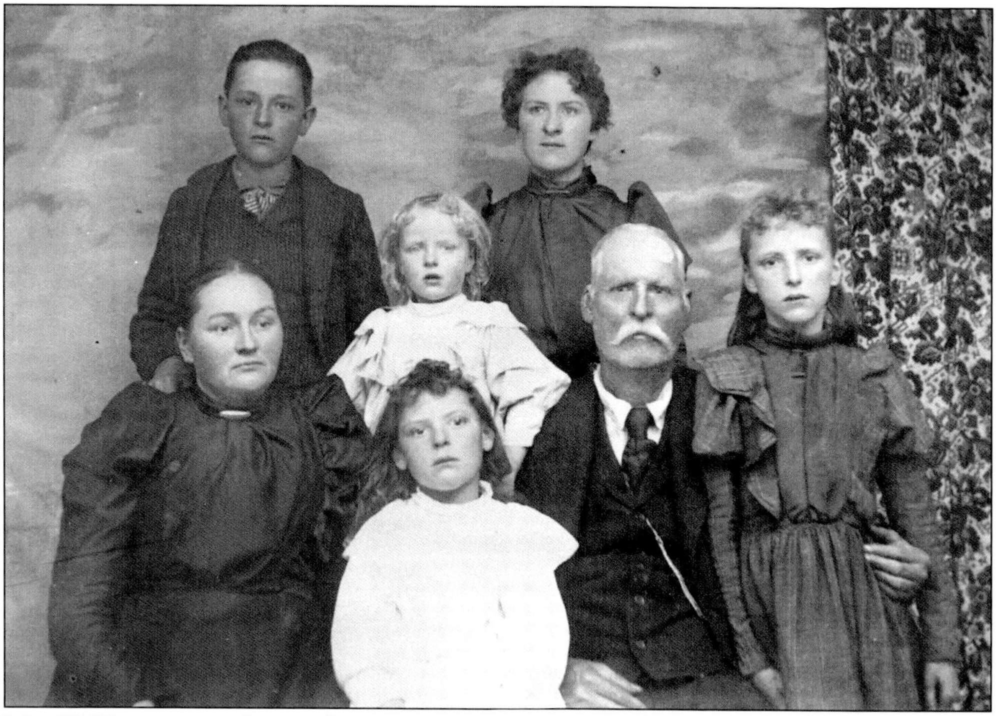

John W. Weaver was an honored pioneer citizen. Born in 1832, he traveled from Missouri with his father, William Weaver, and James Weaver. Arriving with them was Henry Adams and his family. In the spring of 1851, they took Donation Land Claims along the South Umpqua River. Weaver engaged in farming and raising stock and grew fruit extensively, having 100 acres in orchards. Pictured above from left to right are (first row) Hallie Weaver; (second row) Clara (Sumner) Weaver, Faye Weaver, John W. Weaver, and Bertha Weaver; (third row, standing) Fred Weaver and Lulu Weaver. In the c. 1910 photograph below, taken on the Weaver place three miles south of Myrtle Creek, are (from left to right) "Jinx" Sumner, Fred Weaver, Hallie Weaver Black, Clara Sumner Weaver, John W. Weaver (holding child), Faye Weaver Baldwin, and unidentified.

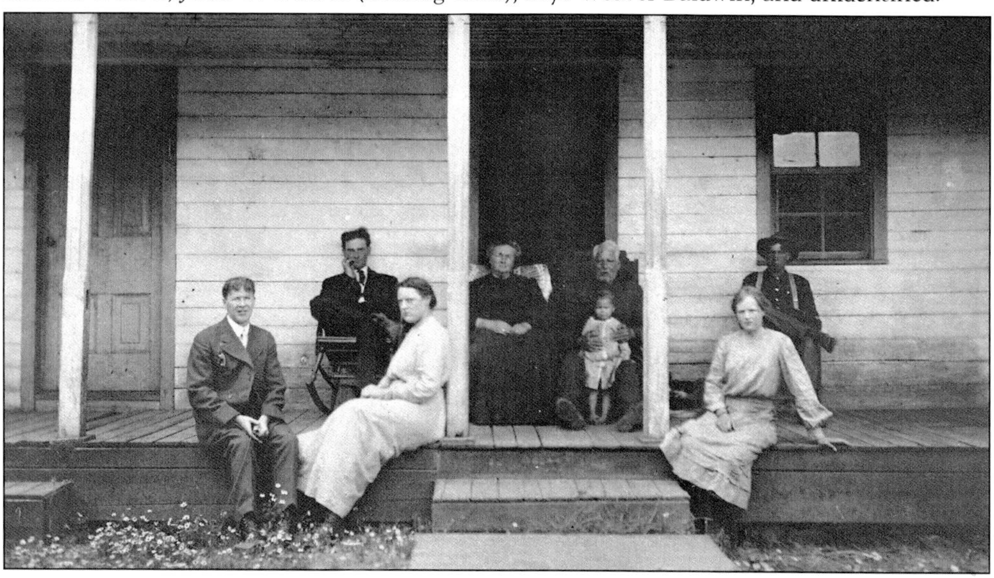

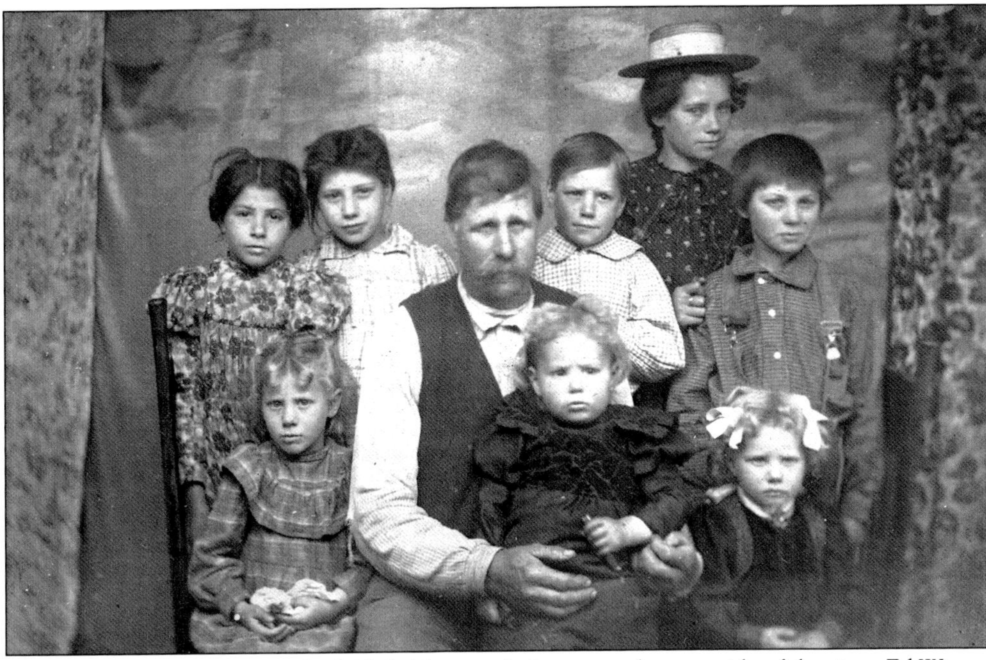

Edwin and Margaret Weaver had 16 children and a home on the west side of the river. Ed Weaver was described in an 1899 issue of the *Plaindealer* as "The most extensive prune grower of Missouri Bottom . . . He has a number of prune orchards which aggregate about . . . 10,000 trees . . . He has been building evaporators for several years but his crop has hitherto increased faster than his evaporator capacity."

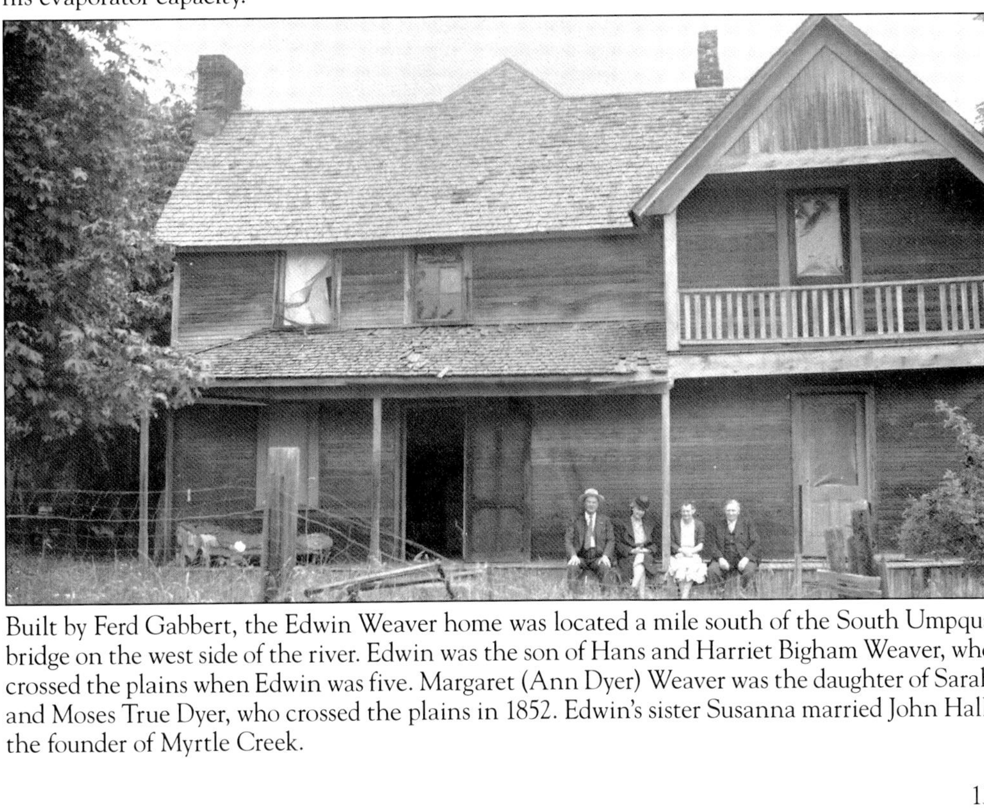

Built by Ferd Gabbert, the Edwin Weaver home was located a mile south of the South Umpqua bridge on the west side of the river. Edwin was the son of Hans and Harriet Bigham Weaver, who crossed the plains when Edwin was five. Margaret (Ann Dyer) Weaver was the daughter of Sarah and Moses True Dyer, who crossed the plains in 1852. Edwin's sister Susanna married John Hall, the founder of Myrtle Creek.

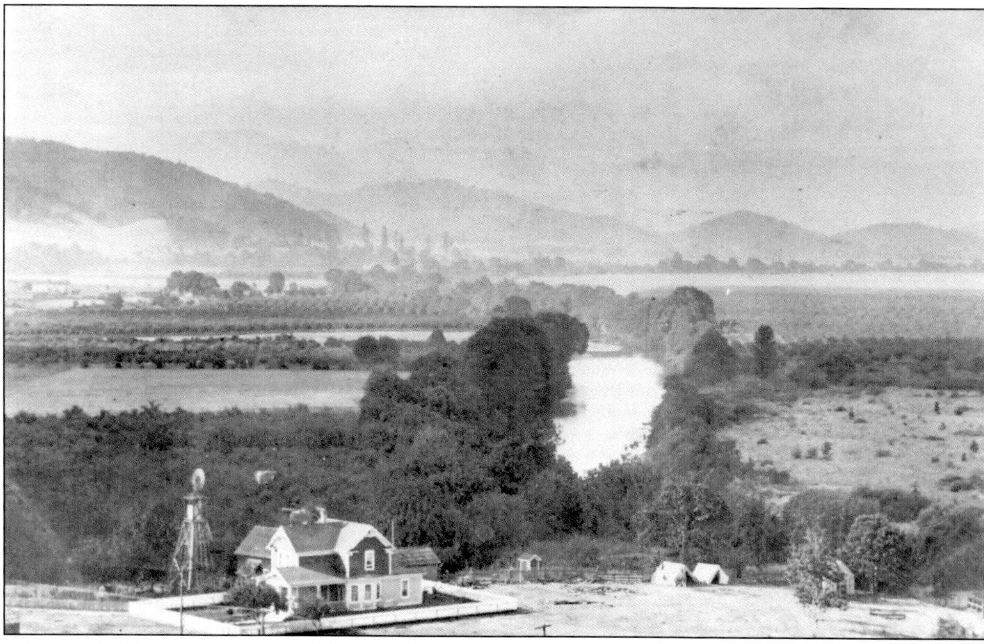

John Hall, Henry Adams, and Hans Weaver were pioneers of the prune industry in Douglas County. This is a view of the Henry Adams home, located one mile south of the gristmill, with acres of prune orchards visible on both sides of the South Umpqua River. Henry Adams, along with William, John, and James Weaver, were the first pioneers to settle here. All being from Missouri, they called this valley Missouri Bottom.

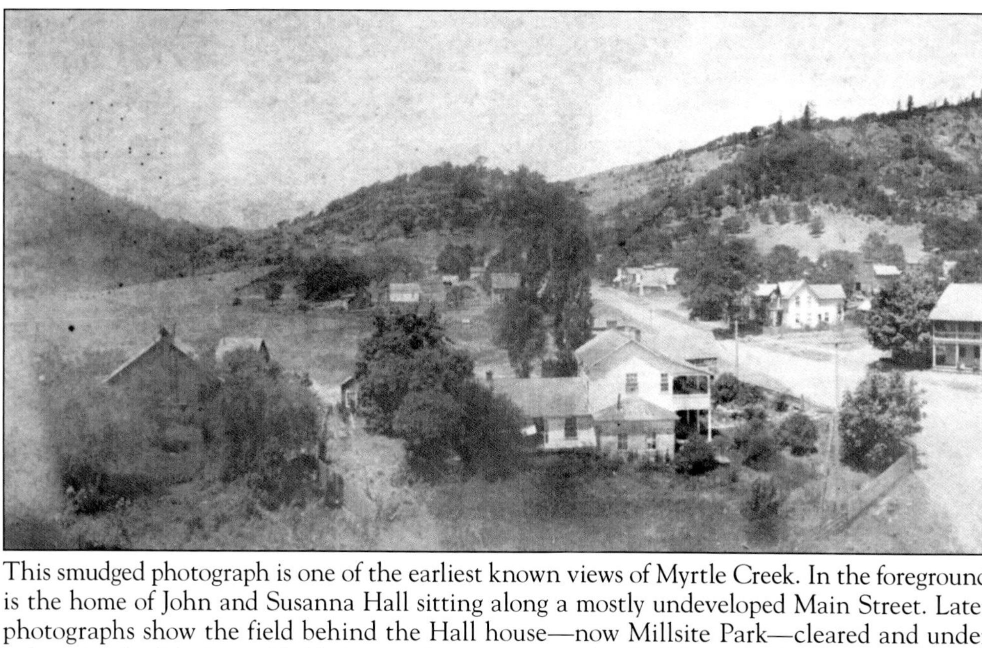

This smudged photograph is one of the earliest known views of Myrtle Creek. In the foreground is the home of John and Susanna Hall sitting along a mostly undeveloped Main Street. Later photographs show the field behind the Hall house—now Millsite Park—cleared and under cultivation by John's son Todd. Across the street from the Hall home is the Overland Hotel. (Courtesy of Alan and Joan Knudtson.)

One of the first Myrtle Creek pioneers, Moses True Dyer, settled six miles up South Myrtle Creek and, around 1853, built the first water-powered lumber mill in Douglas County. The mill was of such importance to early settlers that on the 1856 Territorial Map of Oregon, there are but four locations identified between Roseburg and Jacksonville. Those sites were Round Prairie, Myrtle Creek, the Dyer Mill, and Canyonville.

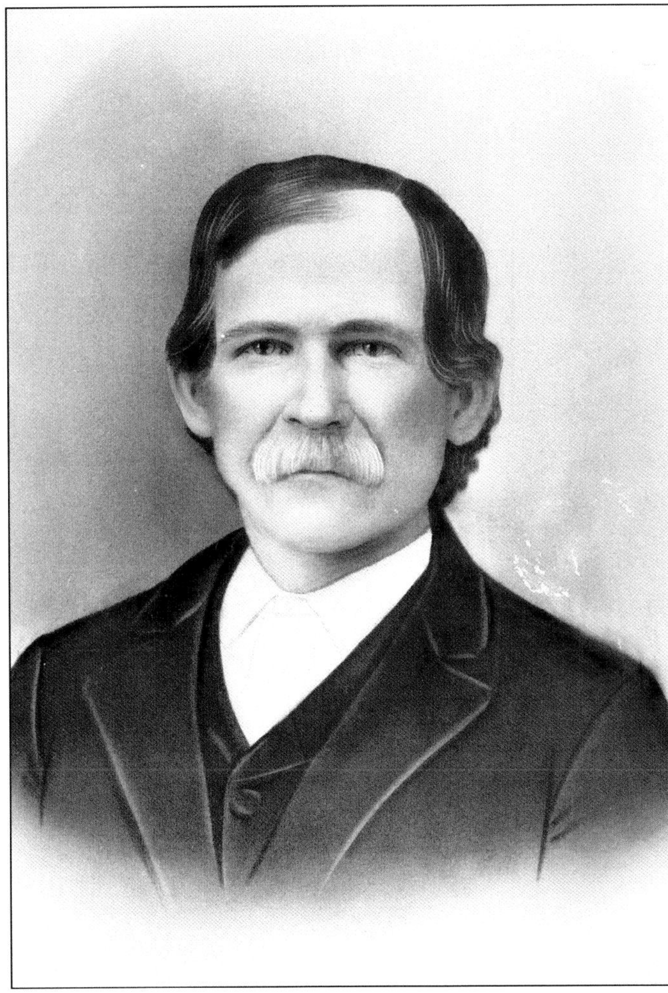

This view shows the Overland Hotel and, to the right of it, the stable used for guests' horses and stock. The little bridge on Main Street near the stable spans the millrace that carried water to power the mill. The water came from a diversion dam up North Myrtle Creek. (Courtesy of Alan and Joan Knudtson.)

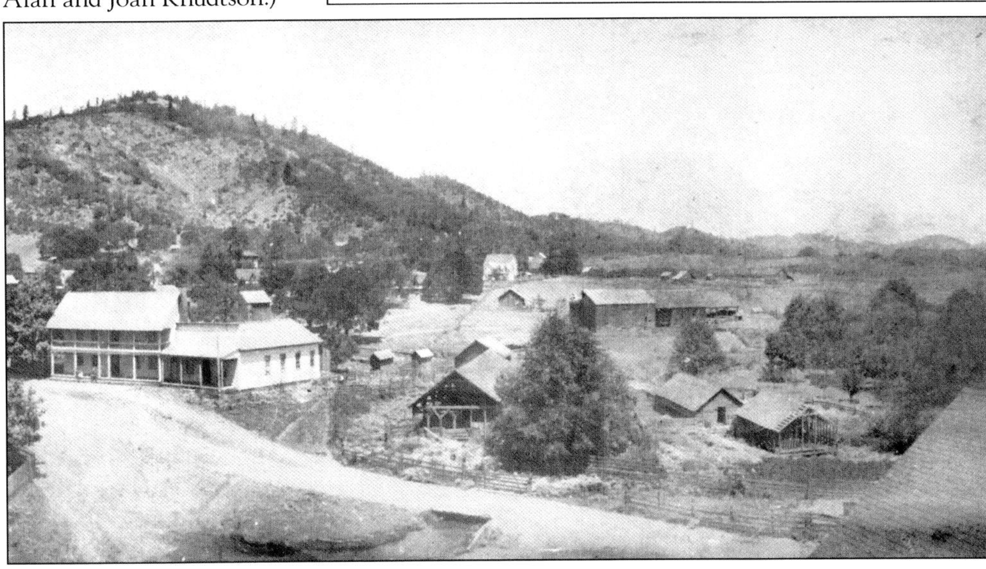

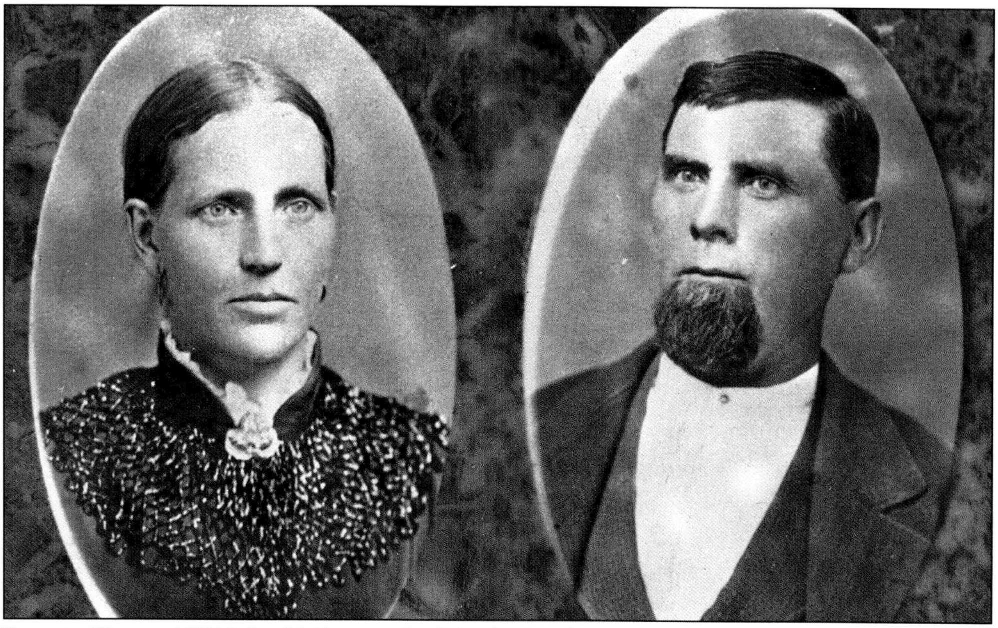

Born in Ohio in 1837, John Hall arrived in Myrtle Creek in 1853. He mined on upper Cow Creek until he purchased half the claim of Lazarus Wright in 1865. Platting the town site in 1868, Hall is considered the founder of Myrtle Creek. Hall was postmaster, county commissioner, president of Citizens State Bank, and a founder of the prune industry. John's wife, Susanna, was the daughter of Hans and Harriet Weaver.

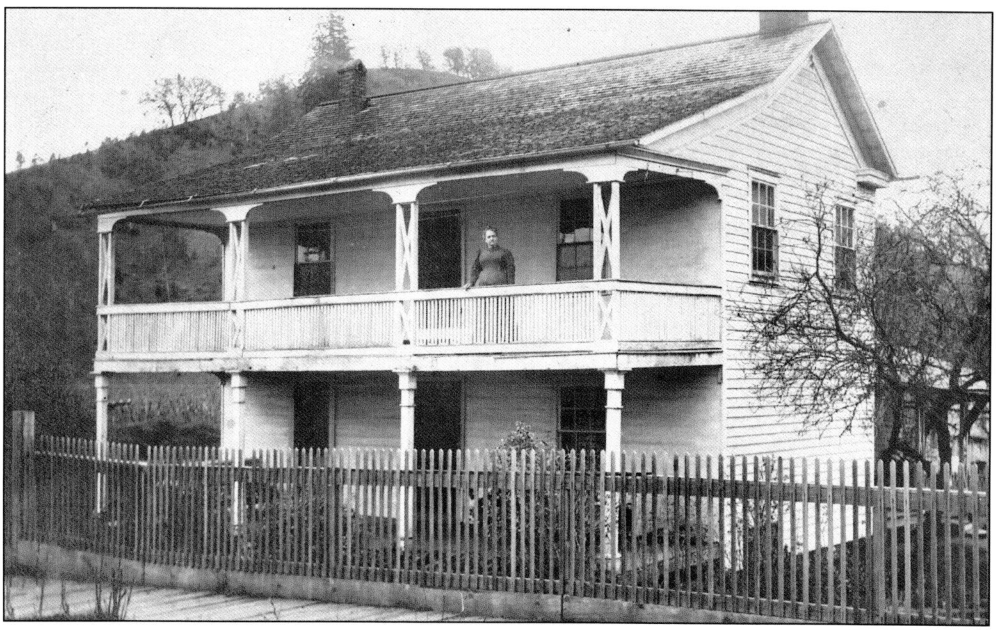

The John and Susanna Hall home, built in the 1860s adjacent to the gristmill on Main Street, was the oldest surviving house in Myrtle Creek when it was demolished in the 1970s and replaced with a parking lot. Susanna Hall fed and boarded many travelers in the house. Standing on the upper porch is the Halls' daughter Emma Hall Buick, who lived in this house her entire life, which spanned nearly 100 years.

The children of John and Susanna Weaver Hall are pictured around 1878. Seated from left to right are (first row) Emily (9), Harriet (11) holding Grace (1), and Florence (7); (second row) John "Todd" (4), William (14), and James (12). Todd later farmed the land now occupied by Millsite Park. In 1905, Florence married Dr. A.C. Seely and moved to Roseburg.

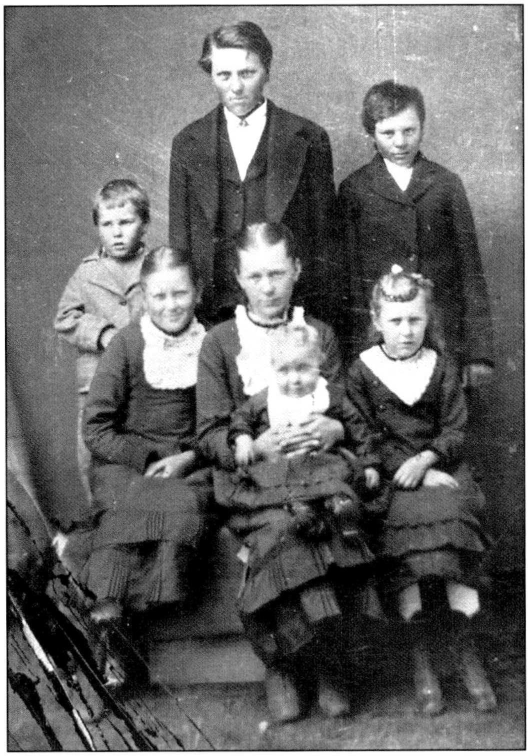

In this photograph, taken by Dr. A.C. Seely in November 1902, John Hall and his second wife, Florence, relax in their parlor in the company of friend and fellow Myrtle Creek pioneer James Weaver. John was a widower, and in 1897, he married Florence, the widow of his brother Jasper.

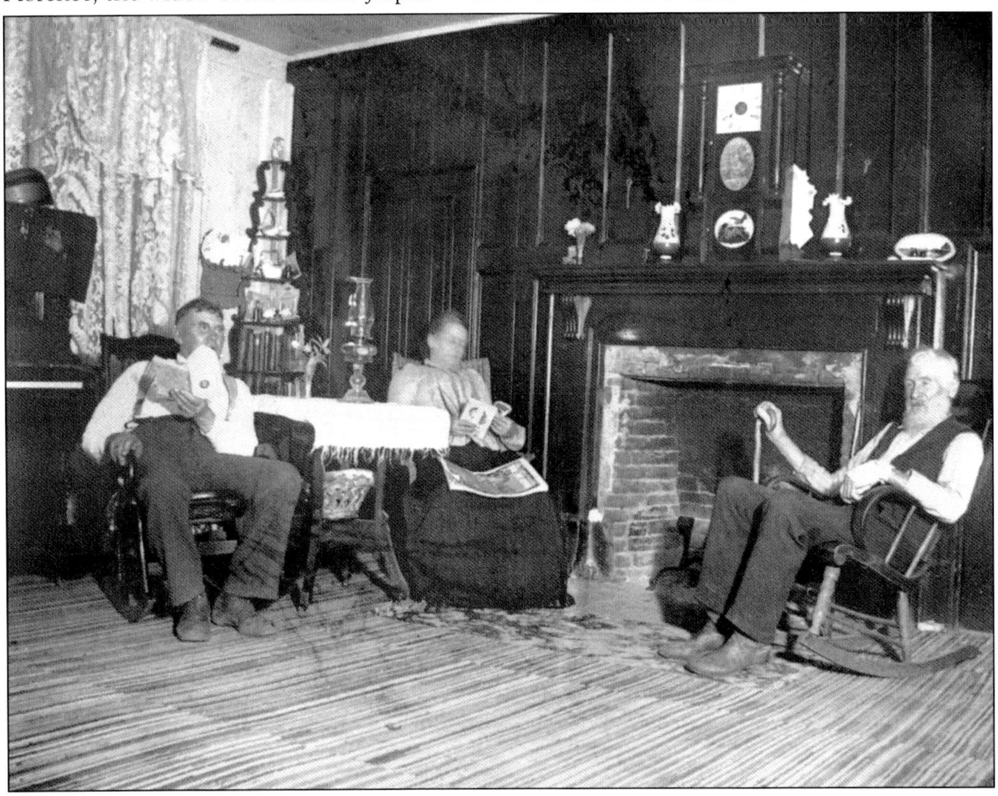

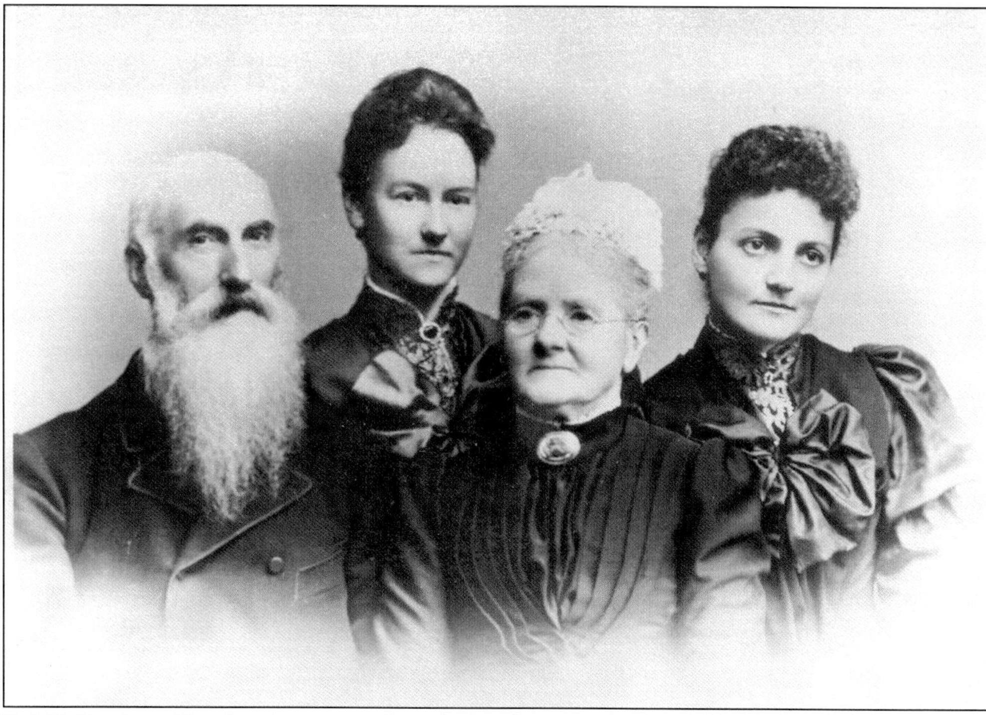

D.S.K. Buick and his family came to Myrtle Creek in 1876 to operate the Overland Hotel. The family became prominent citizens, filling functions from riding shotgun on the stage to running the telegraph office. They moved to Germany in 1891 after being appointed by President Harrison to the position of German consulate. A son, Charles, married to John Hall's daughter Emma, remained in Myrtle Creek. (Courtesy of Canyonville Pioneer-Indian Museum.)

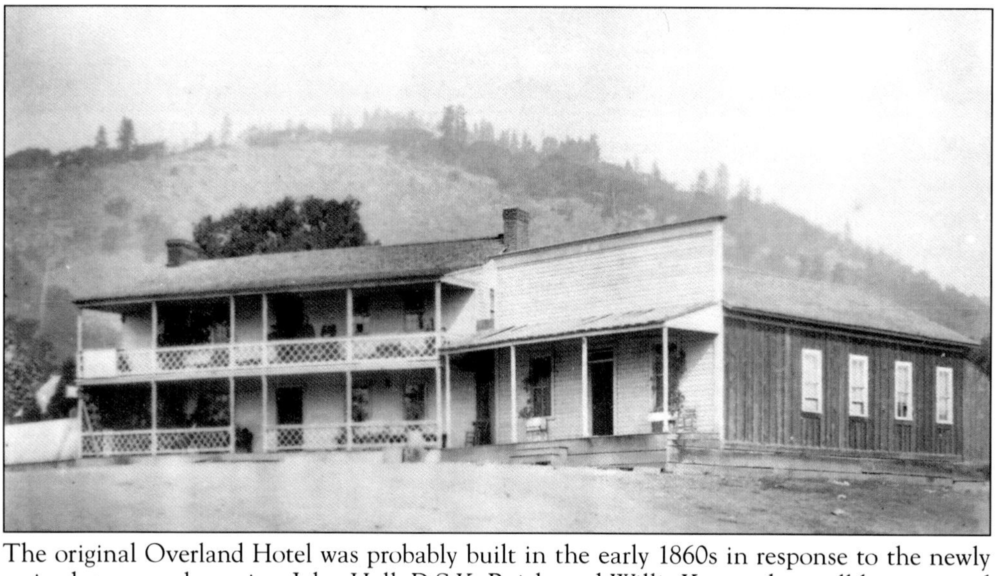

The original Overland Hotel was probably built in the early 1860s in response to the newly arrived stagecoach service. John Hall, D.S.K. Buick, and Willis Kramer have all been named as proprietors. Pictured here is the new Overland Hotel, built around 1890 on the north side of Main Street. It is a fixture in the oldest photographs of Myrtle Creek. In 1905, the hotel burned down and was not rebuilt.

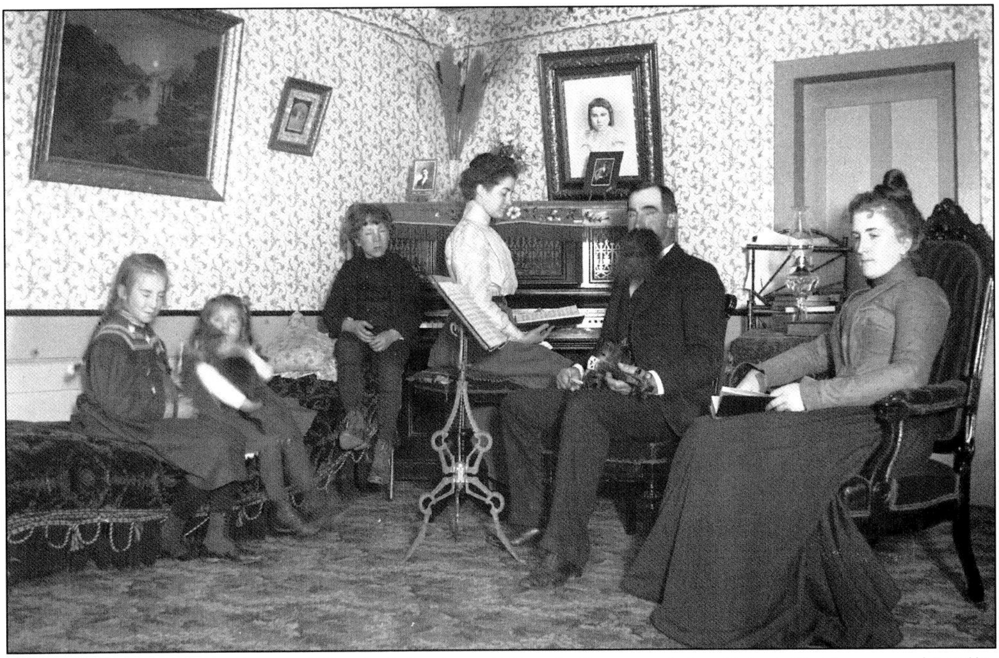

Seated from left to right are Ella, Mildred, and Frank Kramer, Miss Ethel Gabbert, and Willis Kramer and his wife, Lulu Gabbert Kramer. Hon. Willis Kramer was elected to the state legislature in 1902. A one-time proprietor of the Overland Hotel, he also owned the gristmill and attached planer mill and invested in mining, owning mines around the state. Their home, now known as the Gabbert house, still stands on First Avenue.

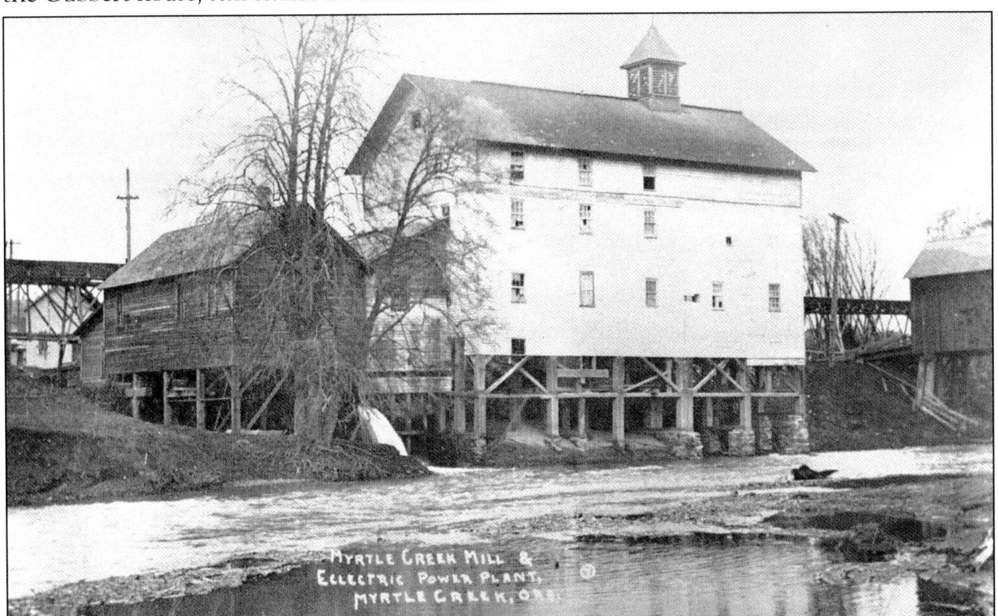

Built by Lazarus Wright around 1856, the Myrtle Creek gristmill had a capacity of 45 barrels a day. The building to the left served as a planing mill and later as an electric power plant. Both were powered by a millrace that ran from a dam up North Myrtle Creek and under Main Street. The outflow of the millrace can be seen between the buildings.

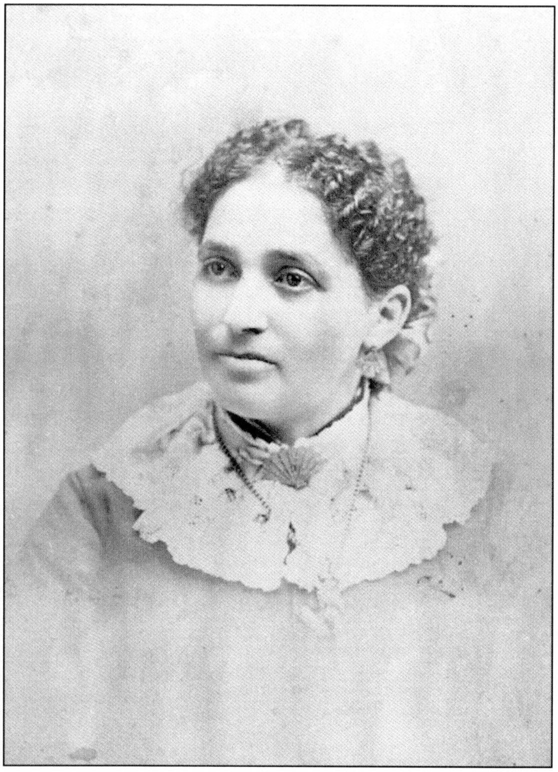

Simon Selig, a mercantile store manager, came to Oregon from Prussia in 1863. He married Helene Solomon (left) in 1868. This union produced seven children. He later moved to Myrtle Creek to take over management of the Marks-Wollenberg store and also became the postmaster of Myrtle Creek. Simon passed away in 1893. The home of Helene Selig (below) was built around 1900 on the corner of Third Avenue and Pleasant Street. Their son Nathan became town mayor in 1905. Nathan owned mercantile stores in Myrtle Creek and Canyonville. Over the years, this home has been enjoyed by Cordelia Chapin Rice, the Ireland family, the Agee family, and, presently, Dan and Wendi Jocoy. Following in the steps of the Selig family, Dan Jocoy is the current town mayor. (Left, courtesy of Douglas County Museum; below, courtesy of Dale Greenley.)

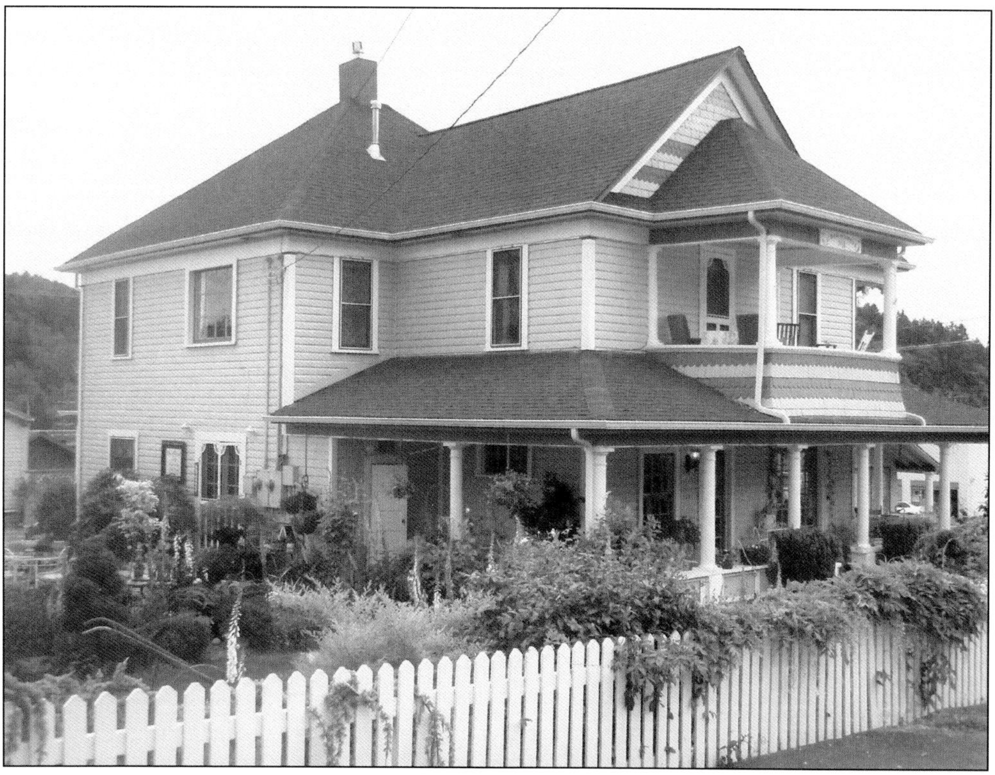

Moses Rice, a pioneer who let no moss grow under his feet, finally settled in Myrtle Creek around 1890. His descendants were movers and shakers who left their marks on Myrtle Creek. According to an old newspaper article, a black walnut tree Moses planted in the 1860s in Oakland later provided some of the "finest furniture veneer ever secured" and was used to construct Governor Patterson's state capital building furniture.

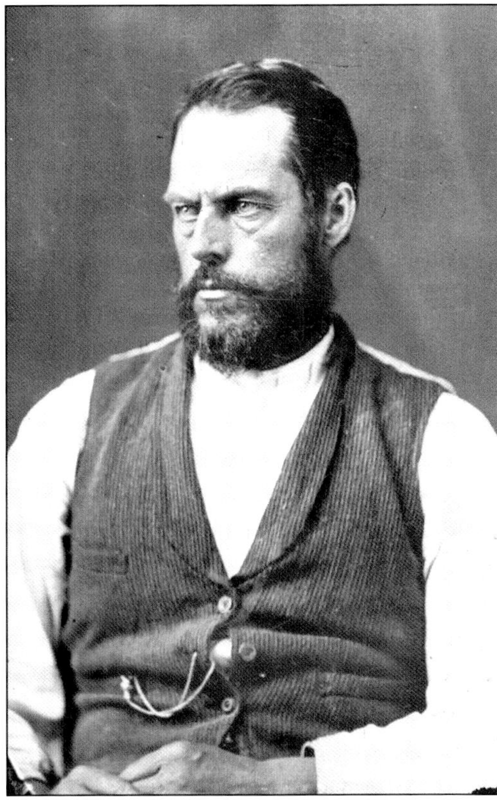

On the front porch of the Hezikiah Dyer house, located on the west side of Main Street near Fourth Avenue, are his wife, Ruby, and children Capitola, Bessie, and Joe. Next to the house, Hezikiah Dyer (a son of Moses True Dyer) operated a store that had previously been the Hunsaker store. The house burned down around 1918, so they remodeled the woodshed and made it their home.

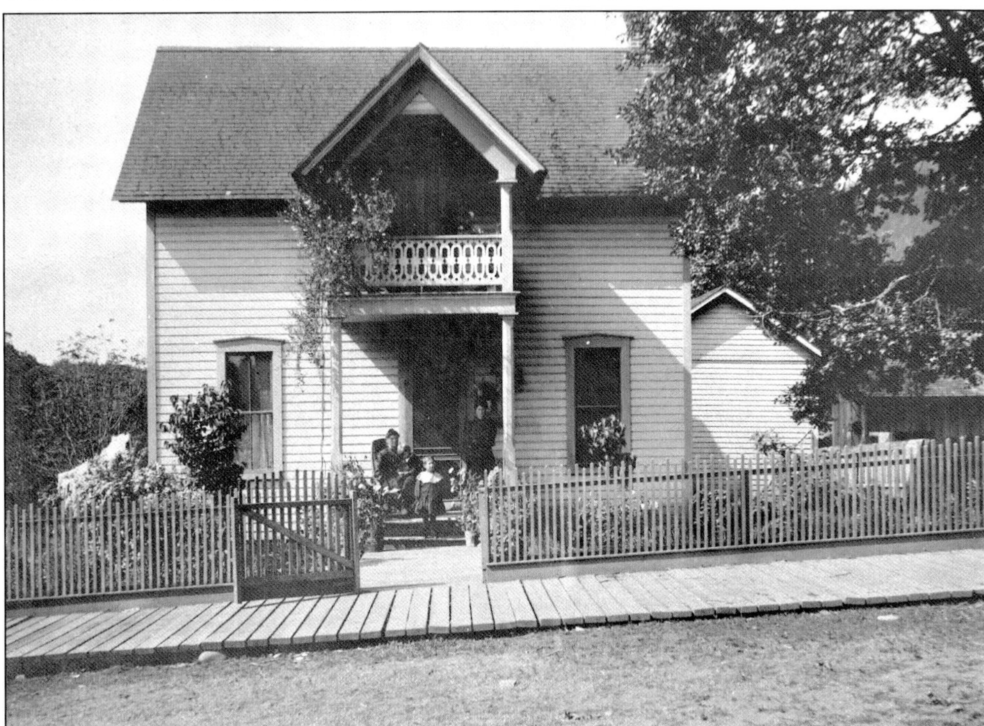

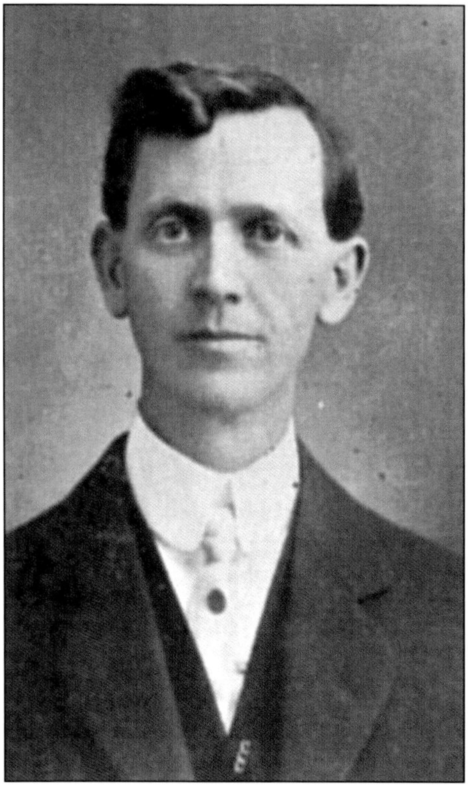

Thurman Chaney had careers as a teacher, principal, author, and superintendent of schools. With Charles Rice in 1897, he started the town's first newspaper, the *Myrtle Creek Beacon*. The *Beacon* failed, but in 1902, Chaney and Rice began publishing the *Myrtle Creek Mail*. In 1911, Chaney wrote a textbook titled *Civil Government*, a copy of which is in the Douglas County Museum.

Posing in front of the *Myrtle Creek Mail* printing office is editor/owner Charles Rice. Started in 1902 with partner Thurman Chaney, the newspaper has had a few name changes but is still in print 110 years later. The only time it was not printed was during World War II. Currently named the *Douglas County Mail*, it is owned and edited by Bob Chaney, a relative of the original owner.

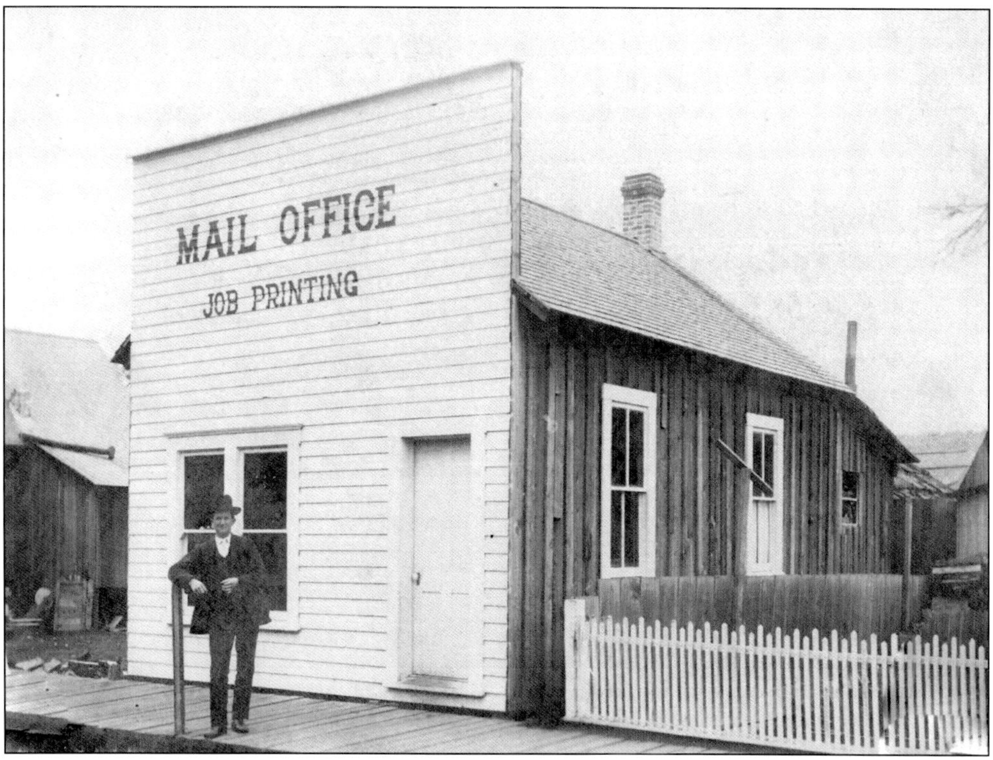

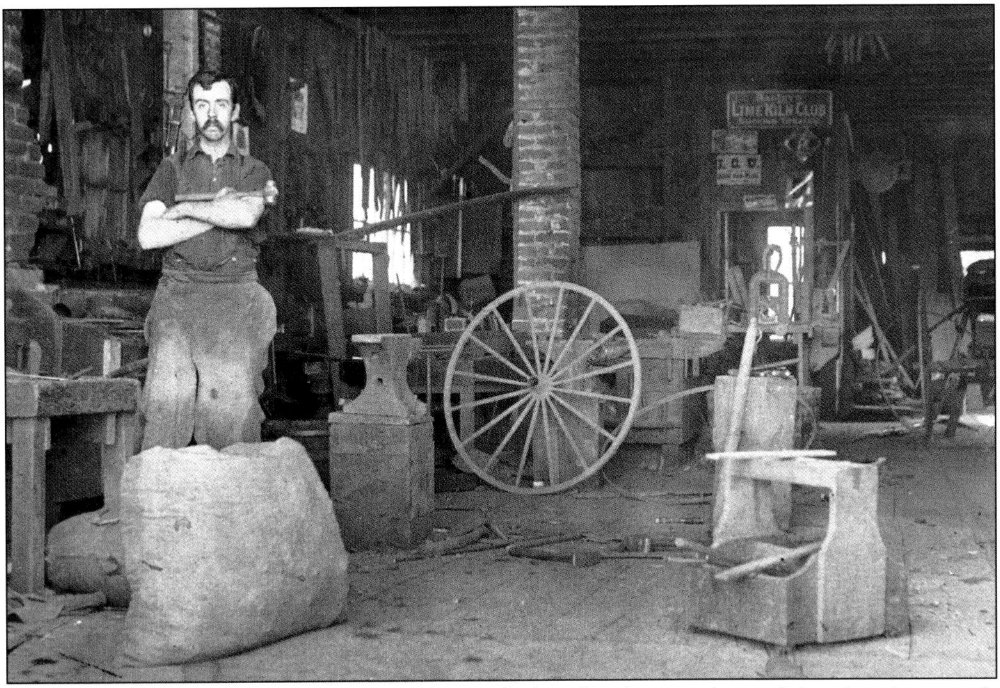

Local lore varies about who this blacksmith is and where his shop was located, but this photograph has some interesting features. On the back wall is a Lime Kiln Club sign, the floor is strewn with tools, and the "smitty" poses beside his anvil with a tool carrier in the right foreground.

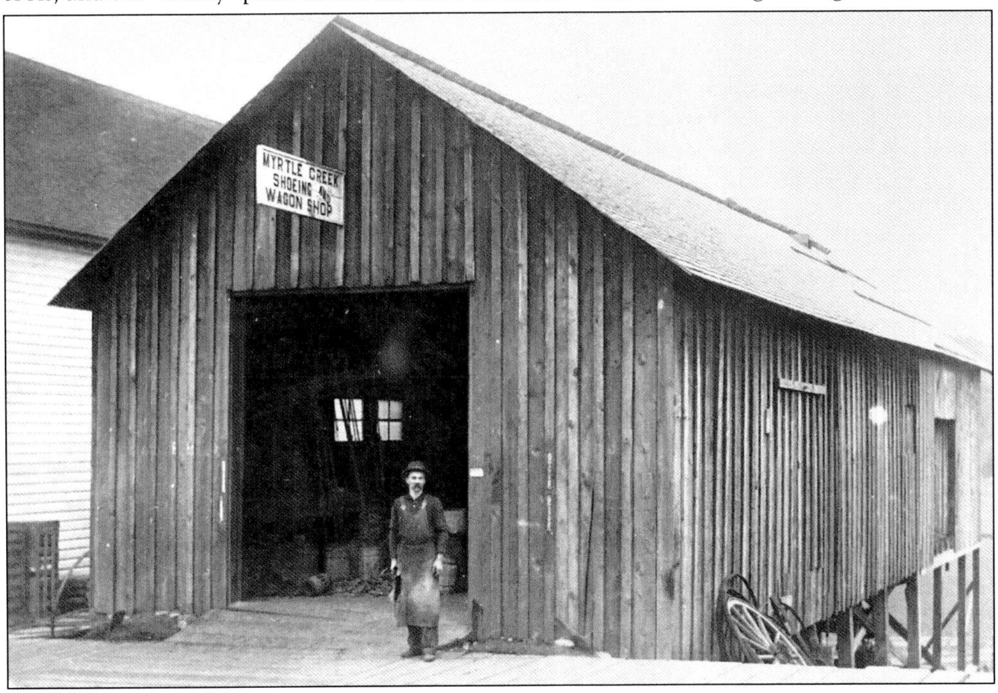

The Myrtle Creek Shoeing and Wagon Shop once stood on stilts at the southwest corner of Main Street and Fourth Avenue. The old-time equivalent of today's automobile repair shop, it has wagon wheels leaning against its walls, and inside are barrels of horseshoes.

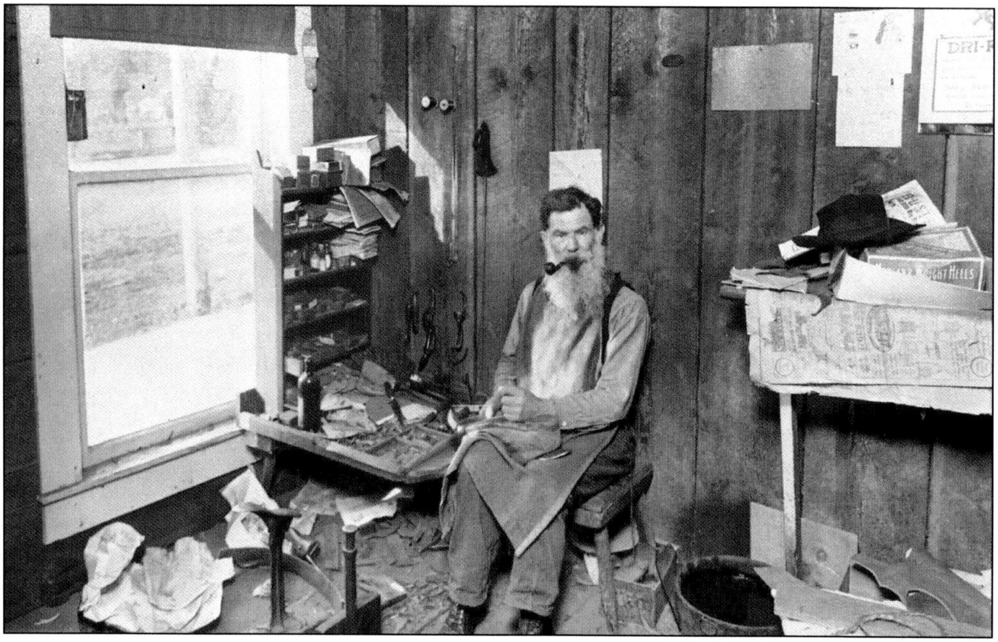

With shoe lasts in front of him and tools hanging on the wall behind, cobbler Dick Harlow plies his trade in his Main Street shop. Harlow lived in a big, white house at the corner of Division and Nichol Streets. His daughter Kate Miles later lived in the same house and though nearly blind, operated a confectionery business in town.

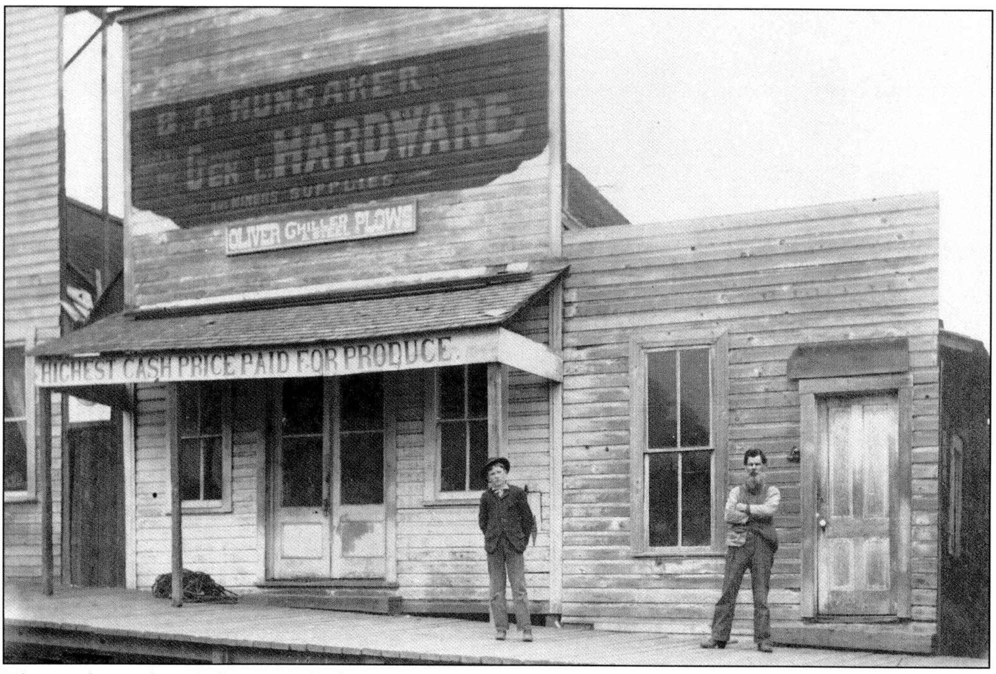

This early, undated photograph shows merchant Ben Hunsaker and cobbler Dick Harlow posing in front of their respective businesses on the east side of Main Street. Hunsaker later built a larger store on the other side of the street (see page 25) and actively served the community for many years.

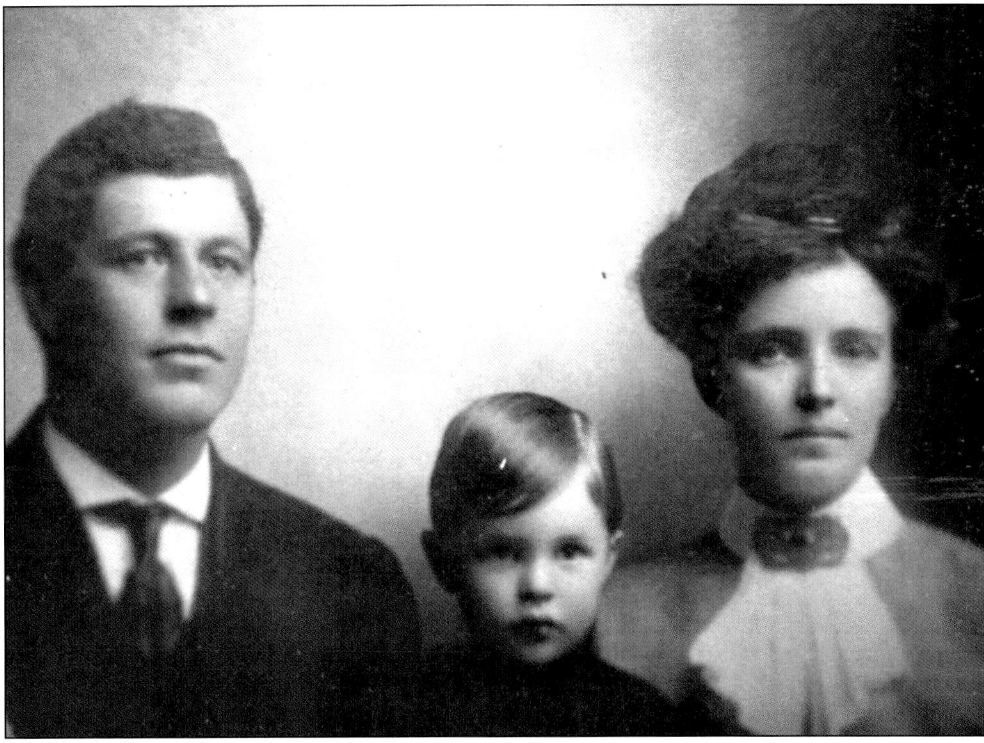

The Hunsaker family had a long business history in Myrtle Creek. Though little is now known about them, their storefronts show up in a nearly 50-year span of Myrtle Creek photographs. In this c. 1910 photograph is Otho, possibly the son of Ben, with his son Raymond and wife, Tempa.

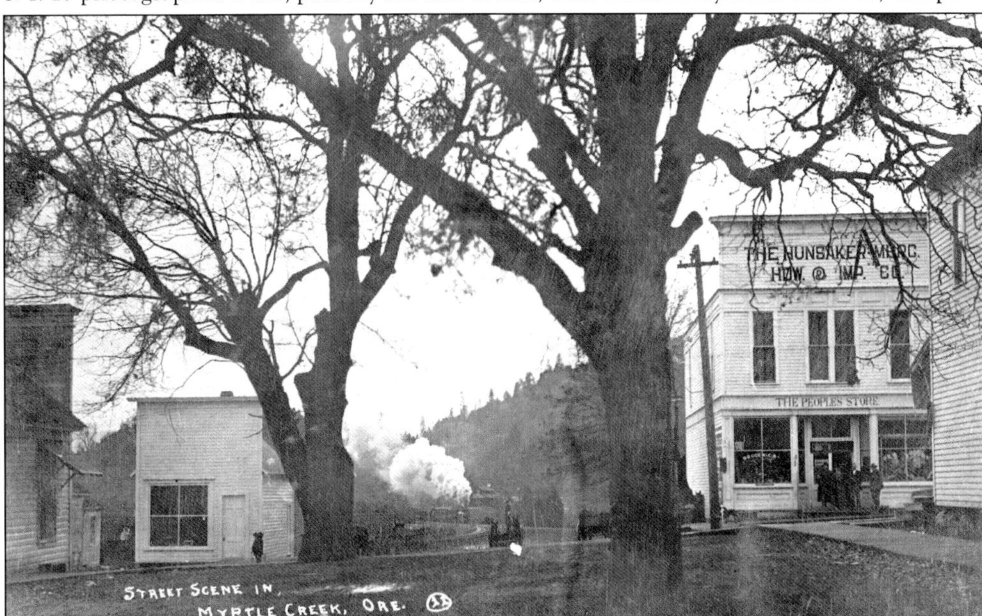

Between John Bryan's watch and jewelry store on the left and Hunsaker's hardware store on the right, the Witness Tree along Main Street and a train leaving the station can be seen. In 1885, John Bryan relocated his store to Roseburg, where it later became the current Knudtson's Jewelers.

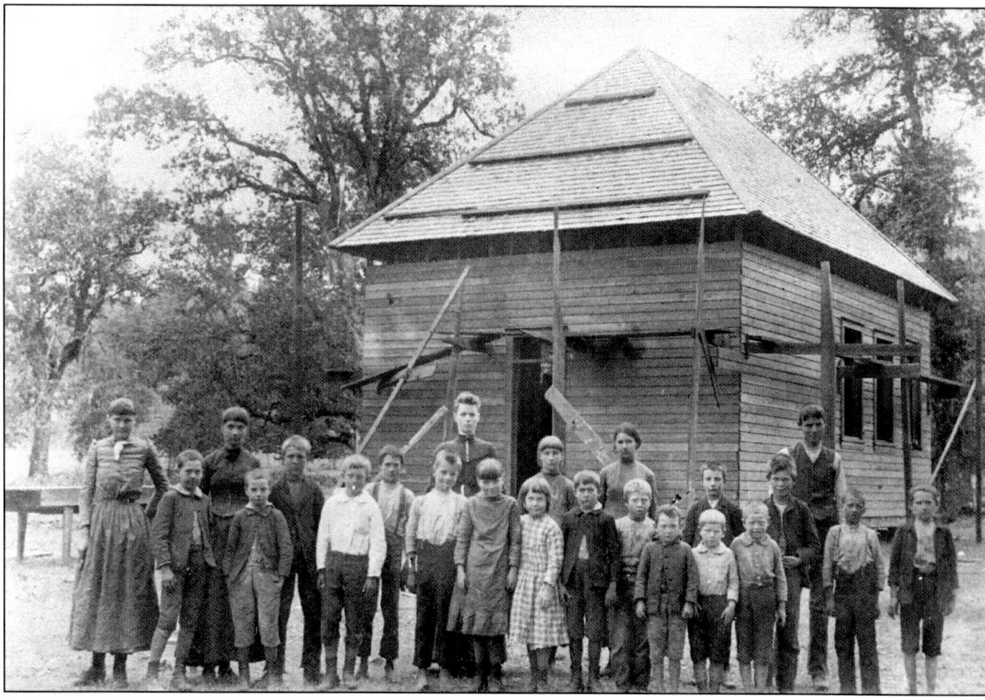

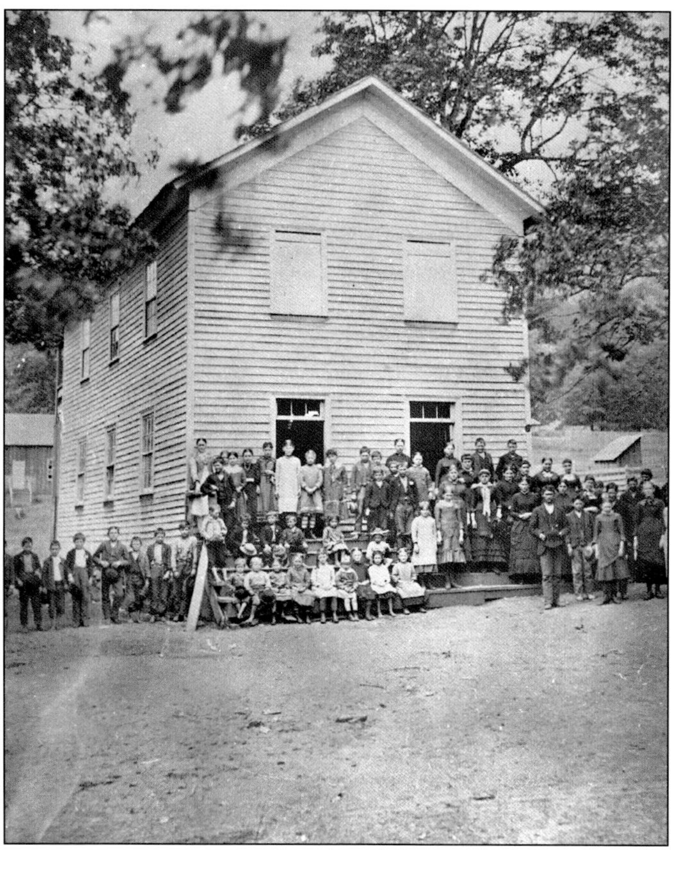

This is possibly the Adams School, located south of town on the Adams Donation Land Claim and just north of the current Norton Lane. The building still stands today with a history of varied uses. If this is the Adams School, then it would be teacher Effie Willis proudly posing with her students in front of the school building, which was still under construction at the time of the photograph.

This large, two-story building, located where city hall now stands, was the first public school in town. From around 1870 to 1890, this building also housed church meetings and Sunday school classes. Around 1890, this roughly built structure was replaced by an attractive new building on the same site.

Two

A Century Turns, A City Grows

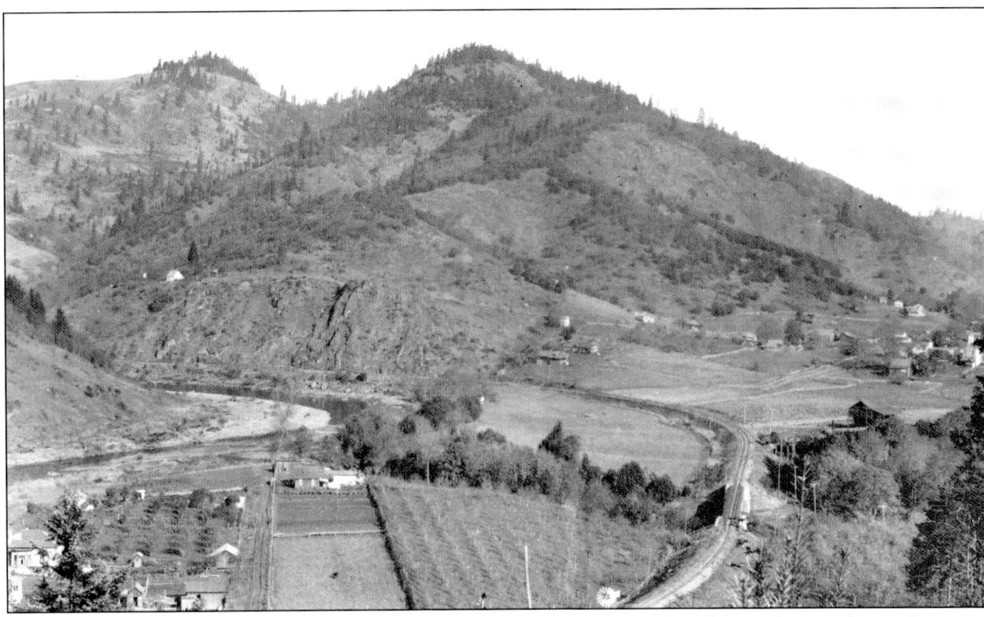

The railroad has arrived! The coming of the railroad in 1882 placed Myrtle Creek on the main Oregon and California line. No longer dependent on the slow, arduous Scott-Applegate/Hooker Military Road to cart people and merchandise in and out of town, the railroad vastly improved Myrtle Creek's shipping and travel capabilities. Rooster Rock (center), derived its name from the ill reputation of the white house on the cliff above the river.

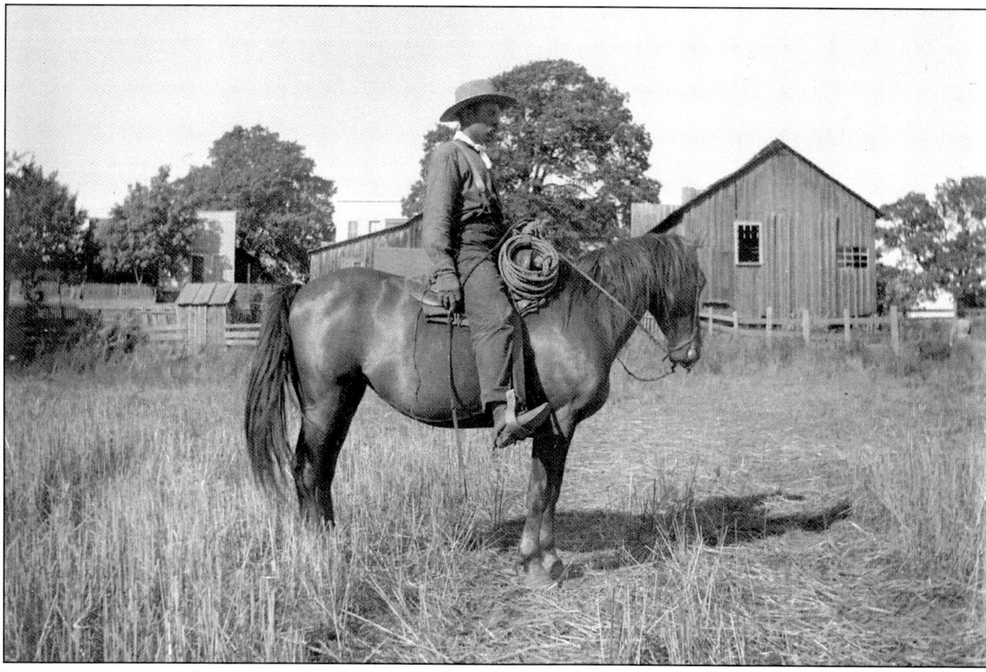

Dr. A.C. Seely took this photograph of John "Todd" Hall in August 1899 in Todd's field. Todd, a son of John Hall, cultivated all of what is now Millsite Park and sits astride his horse Topsy about where the Millsite Park restrooms are now.

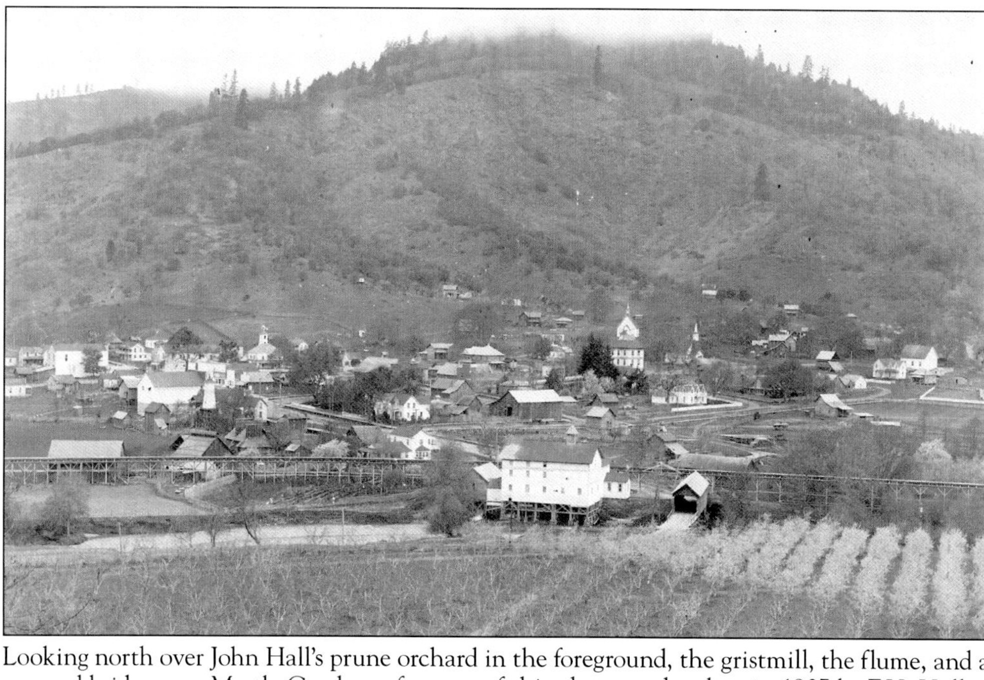

Looking north over John Hall's prune orchard in the foreground, the gristmill, the flume, and a covered bridge over Myrtle Creek are features of this photograph taken in 1907 by F.H. Hull, a Medford photographer. Above the covered bridge is the multiple-gabled Kramer-Gabbert house on First Avenue, the two-story school with the Christian Church beside it, and the Presbyterian Church directly behind.

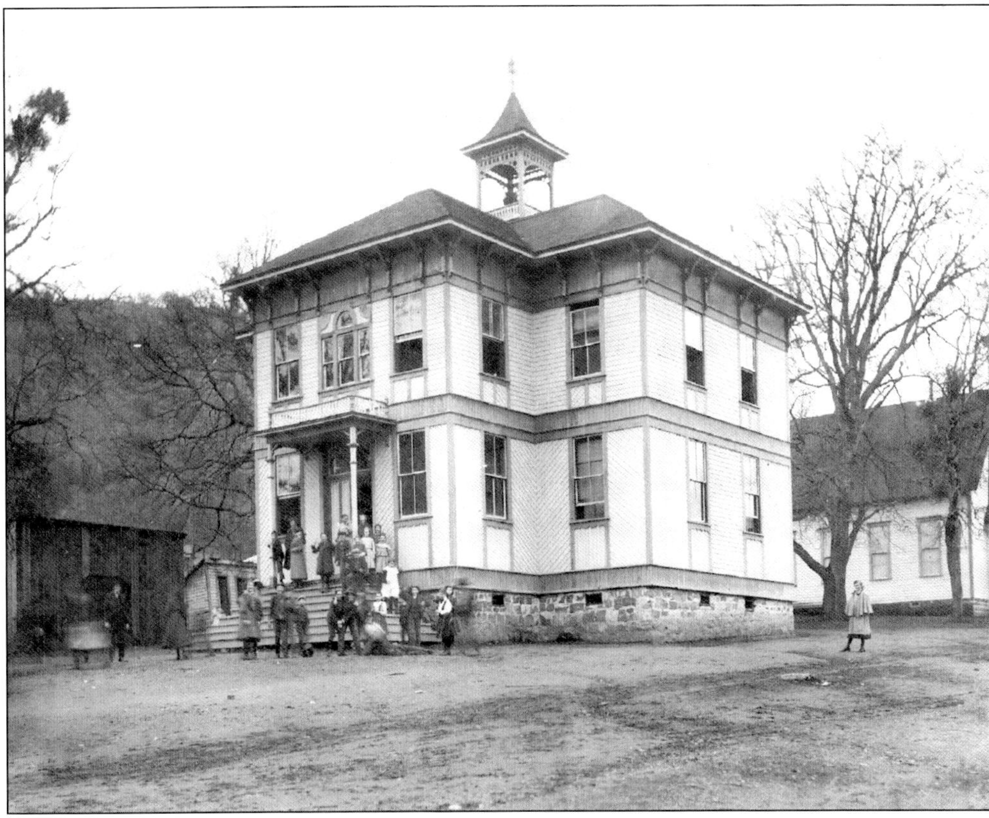

This attractive school building with its bell tower was often pictured on postcards. Built around 1890 on Pleasant Street, the youngsters were on the first floor and the older students occupied the second floor. In 1909, a new school was built on Division Street and the old school was converted to City Hall. This structure served the municipality until it was replaced in the 1950s with the present city hall. (Courtesy of City of Myrtle Creek.)

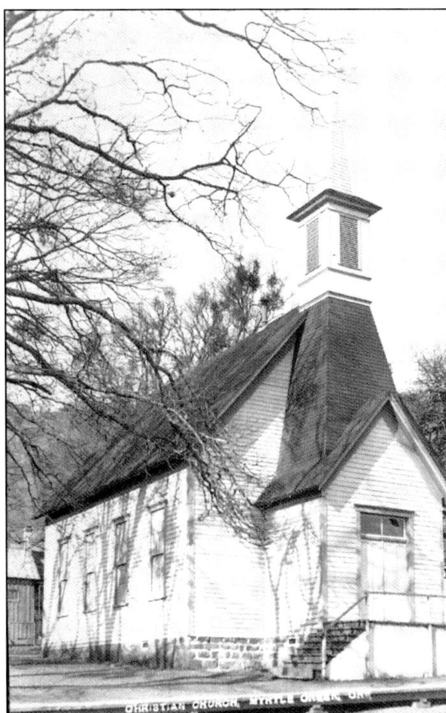

The Christian Church was established in 1855 with Alpheus and Sophia Ireland among the charter members. Meeting at the Pleasant Street School, Dr. Edmund G. Browning, a pioneer minister and physician, preached each third Sunday. Volunteers built the new church shown here, which was completed in 1890 and dedicated by renowned evangelist A.B. Wade. It is one of two historic churches still standing in Myrtle Creek.

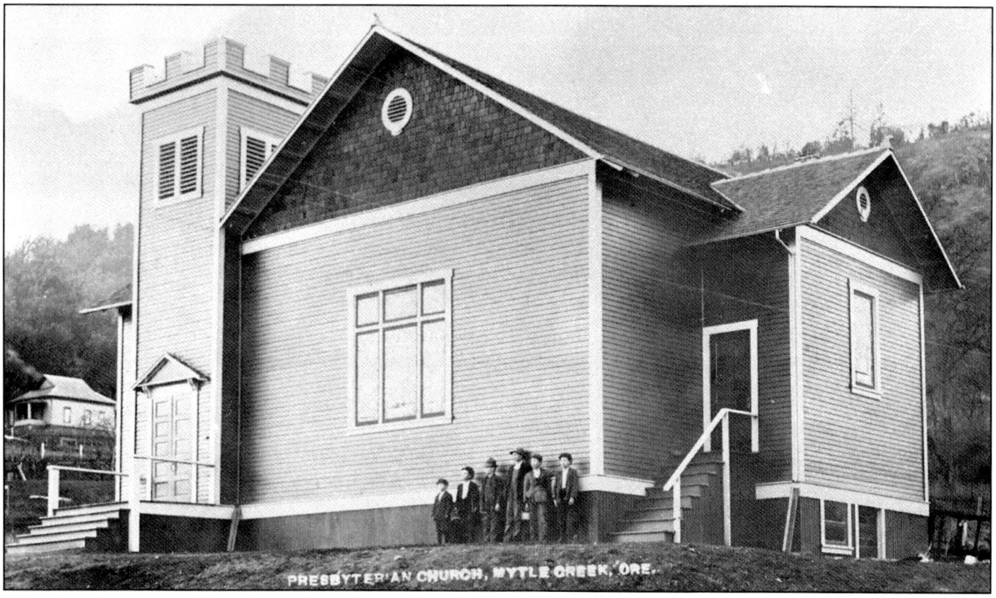

The Presbyterian Church is the white church with the tall steeple, which is clearly seen on the north side of town in nearly all old photographs of the city. That church burned down and was replaced with this attractive building. Services for the Methodist Church were also held here until 1892 or 1893. The building was later purchased by the local grange, which donated the stained-glass windows to the Methodist Church shown below.

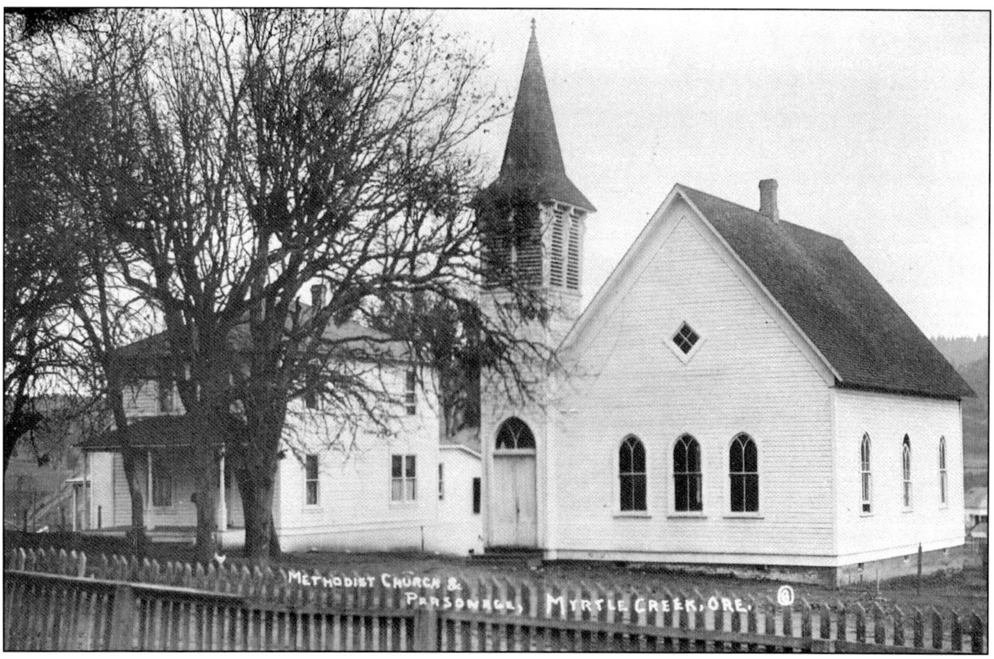

The Methodist Church on Second Avenue was built in 1892–1893, and the adjacent parsonage was built a few years later. The church once suffered an embarrassing accident. Around 1938, minister Jim Wilson cut down the big black oak tree that stood beside the church. He miscalculated, and the tree hit the steeple, knocking it awry. They straightened it as best they could, but to this day, it is still a bit askew.

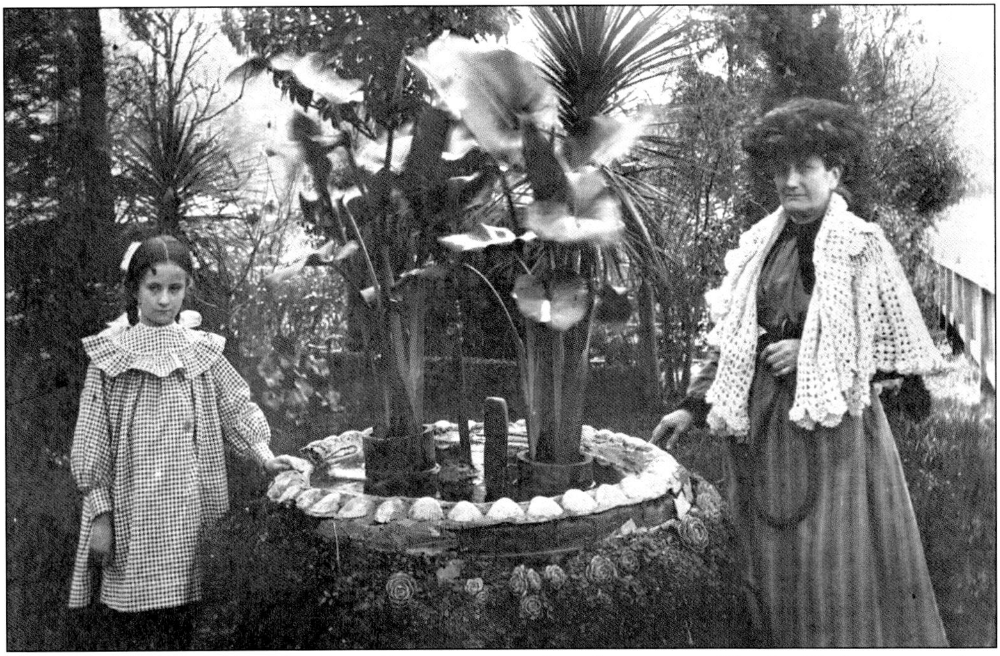

Elizabeth Drake and her daughter Susie pose beside a fountain in Elizabeth's famous garden. Stretching the whole length of the block on the south side of Second Avenue between Pleasant and Oak Streets, the garden included the Magnolia tree she planted that can still be seen there today. She grew rare and unique plants, and her garden was featured in a c. 1900 gardening magazine.

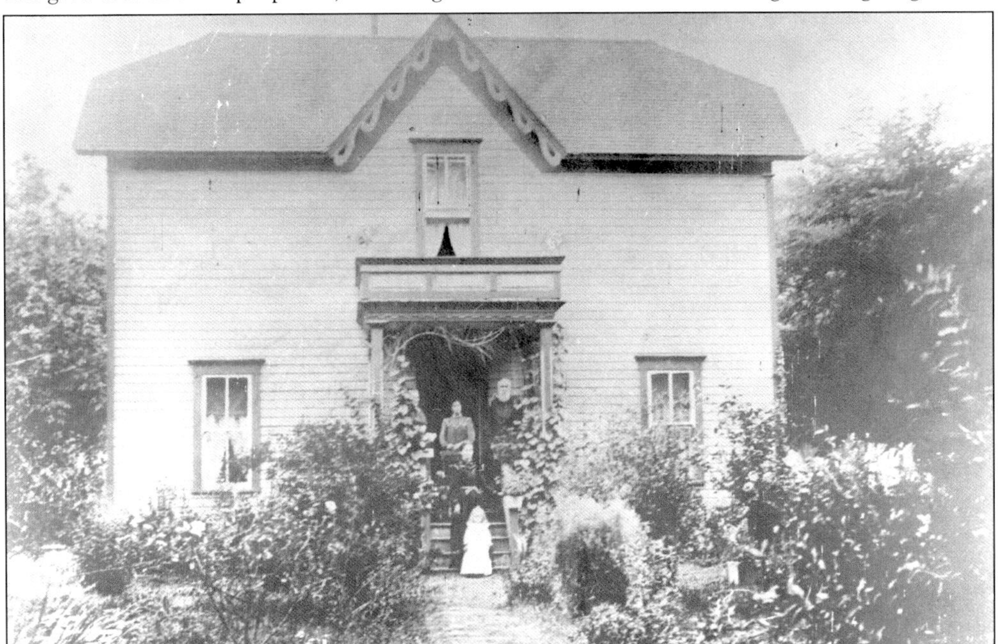

The Drake house, built around 1885 on the south corner of Second Avenue and Oak Street, featured an extensive garden fit for the mayor's house. Mayor William Drake, a Union veteran, was wounded at the battle of Murfreesboro. Later occupied by Madge Gazley, the house was razed in 1974 and replaced with Doctor Goodwin's optometry office.

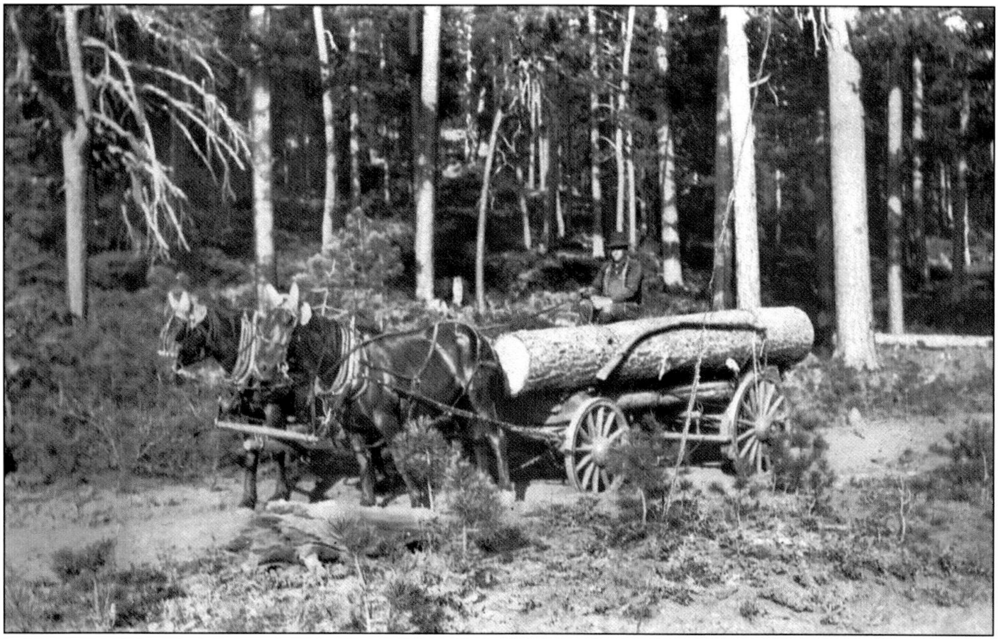

From Moses Dyer's pioneer sawmill to the post–World War II timber boom, logging has been an important part of Myrtle Creek's economy. Getting logs to the mill has always been a challenge. These photographs show two of the means of power used to accomplish the task. (Courtesy of Dorothy Shartner White and Gladys Russell Shartner.)

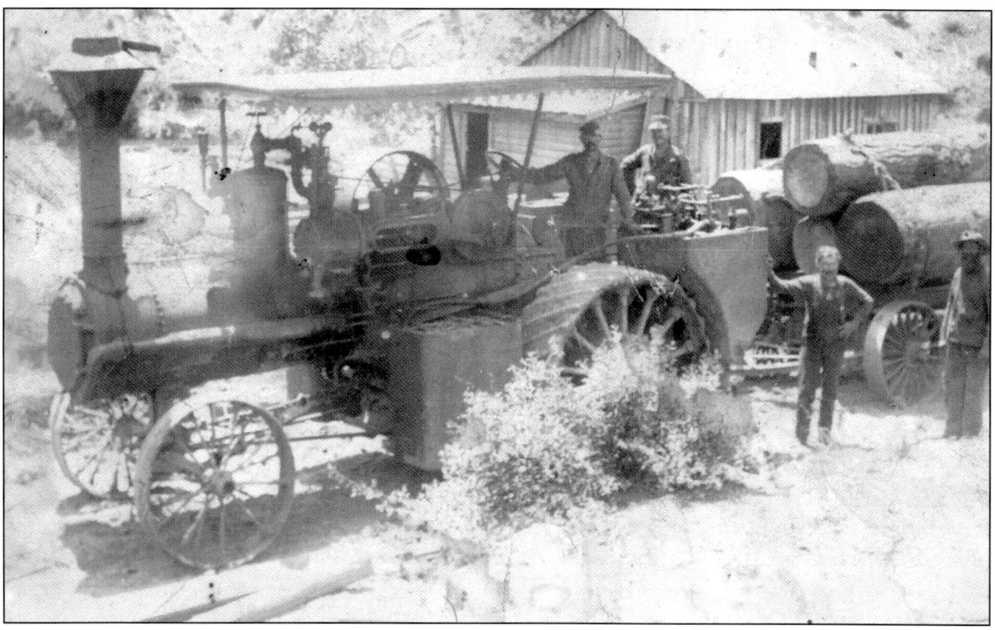

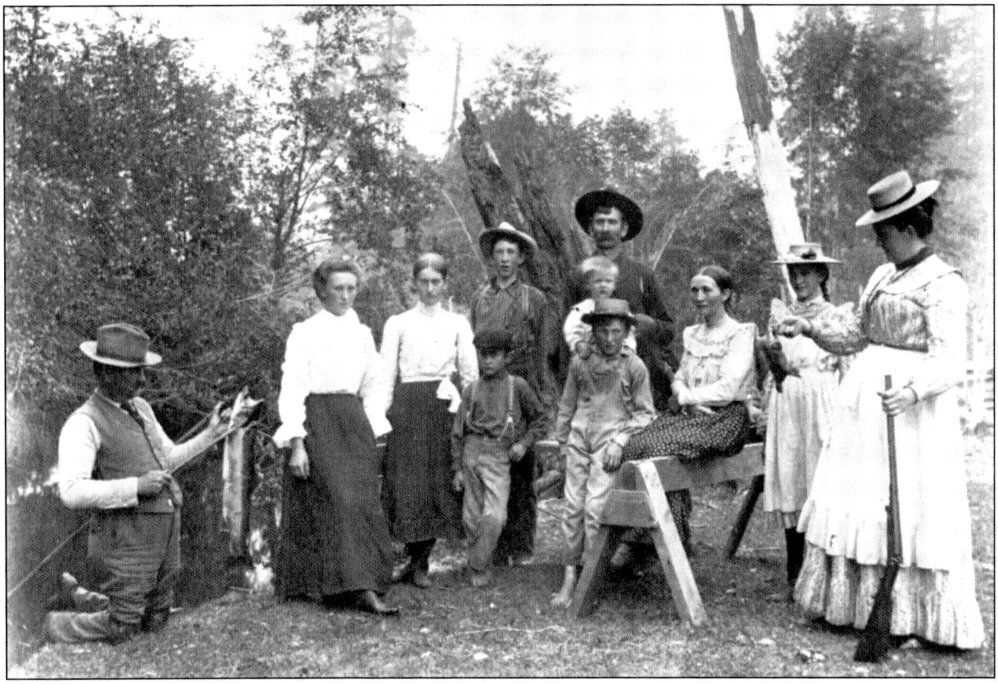

John (left) and Anna Langenberg Bryan flank a group of companions on one of their frequent sporting ventures. A noted outdoorsman who had to walk on his knees, John was featured in a 1903 *Sports Afield* article. Holding a fine salmon while Anna sports a brace of quail, they appear satisfied that his business beginnings at Myrtle Creek in 1878 have brought them success.

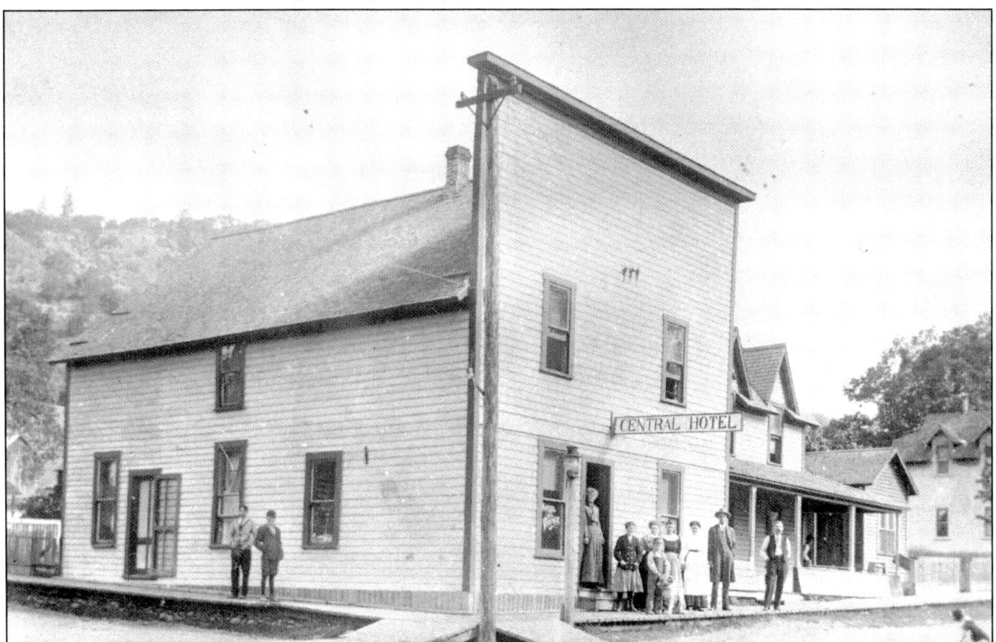

On the site of the current public parking lot at the corner of Second Avenue and Oak Street, there once stood the large Central Hotel. In this turn-of-the-century photograph, a dog watches from the street as several people dressed in their Sunday-go-to-meeting clothes pose for the camera.

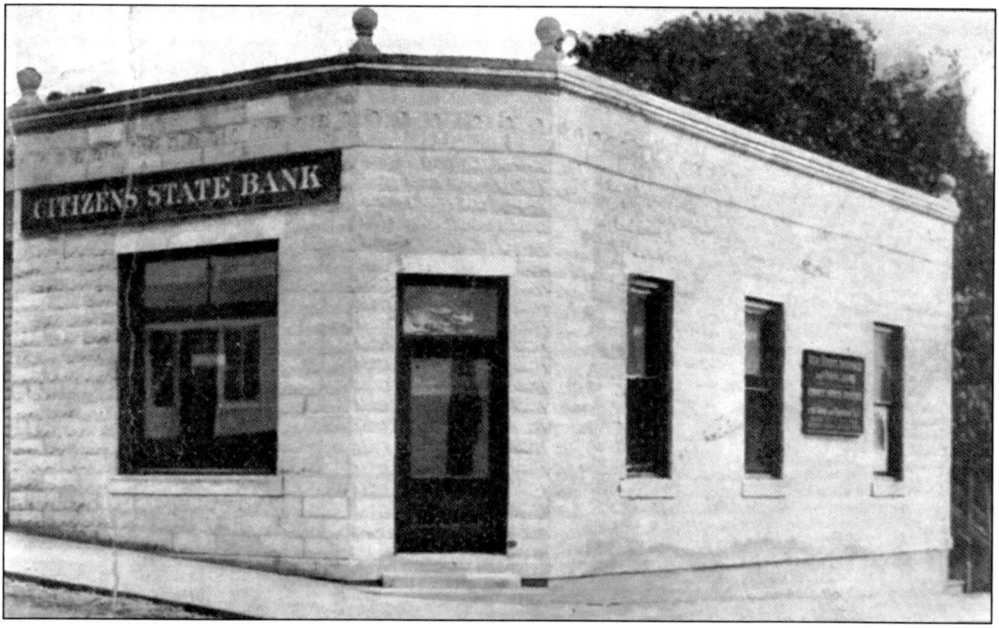

The Citizens State Bank was located at the east corner of Main Street and Second Avenue and is easily recognized in old photographs because of its corner entrance. John Hall, one of the incorporators, also served as the bank president. For many years, Guy Bates served as bank manager. The current Citizens Bank building replaced this one in 1930; when the bank closed, Myrtle Drugstore moved in.

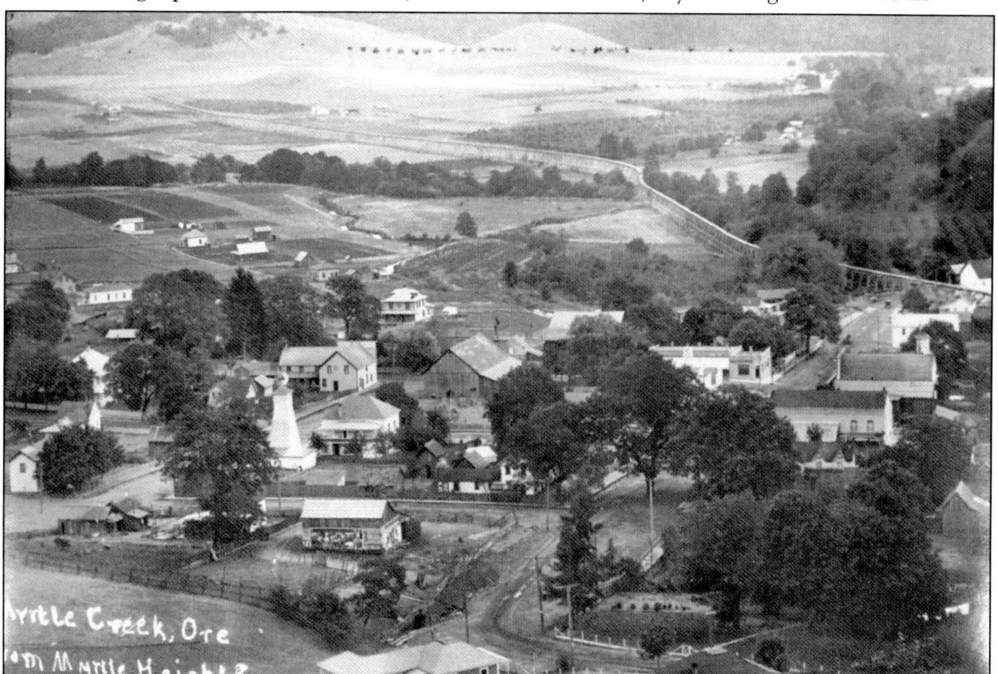

This beautiful panorama of Myrtle Creek, with the graceful curves of a flume sweeping through the background, adorned a 1910 calendar. Built in 1903 using 2.5 million board feet of lumber, the flume cost $25,000 and came from the Johnson Mill six miles up South Myrtle. Reaching a height of 50 feet in places, the flume carried lumber to a planer mill near the railroad tracks.

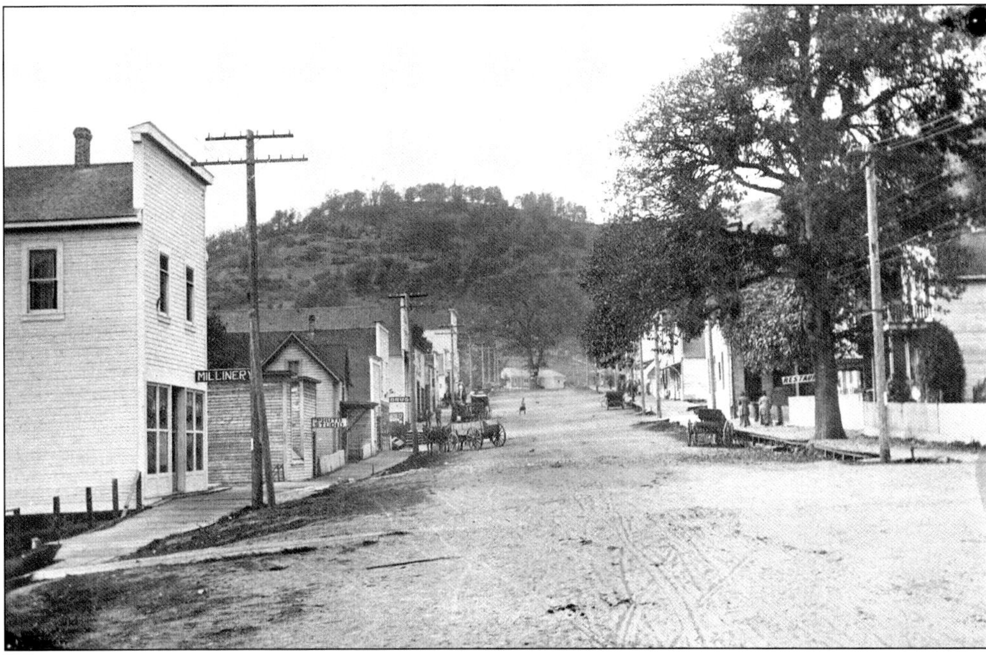

This early 1900s, post-electricity Main Street scene shows a millinery shop, a photograph studio, and a drugstore on the left. On the right is a restaurant, and the street is lined with buggies and wagons. People walking the wooden sidewalks include two ladies on the right in long, flowing dresses.

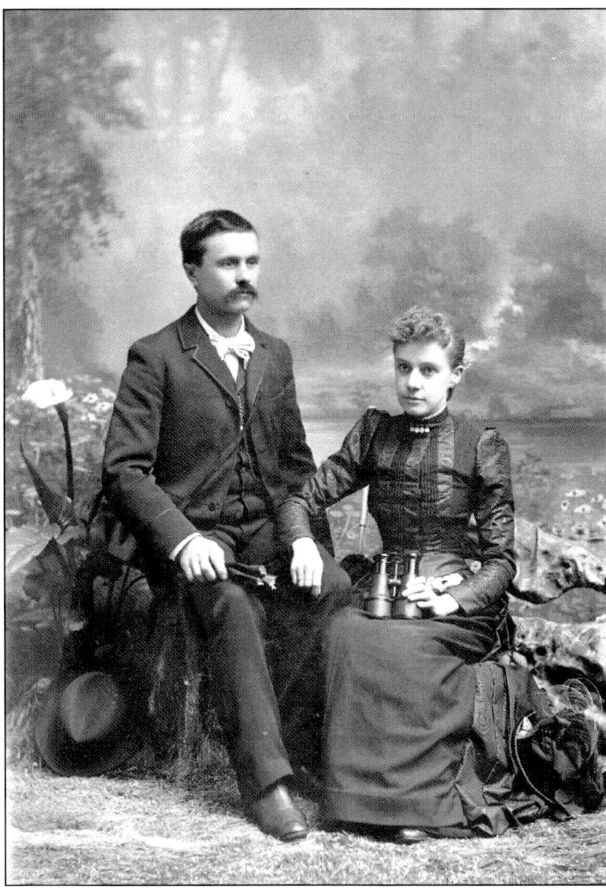

Hubert D. Graves and his wife, Lou Rice Graves, are seen in this c. 1895 photograph. Hubert had a photography studio in Roseburg and took many of the turn-of-the-century images included in these pages. There are indications that they may have had a studio in Myrtle Creek and lived here for a time, but the evidence is not conclusive. Lou Rice Graves is holding a pair of binoculars, but it is a shame that she is not displaying the beautiful hat lying beside her.

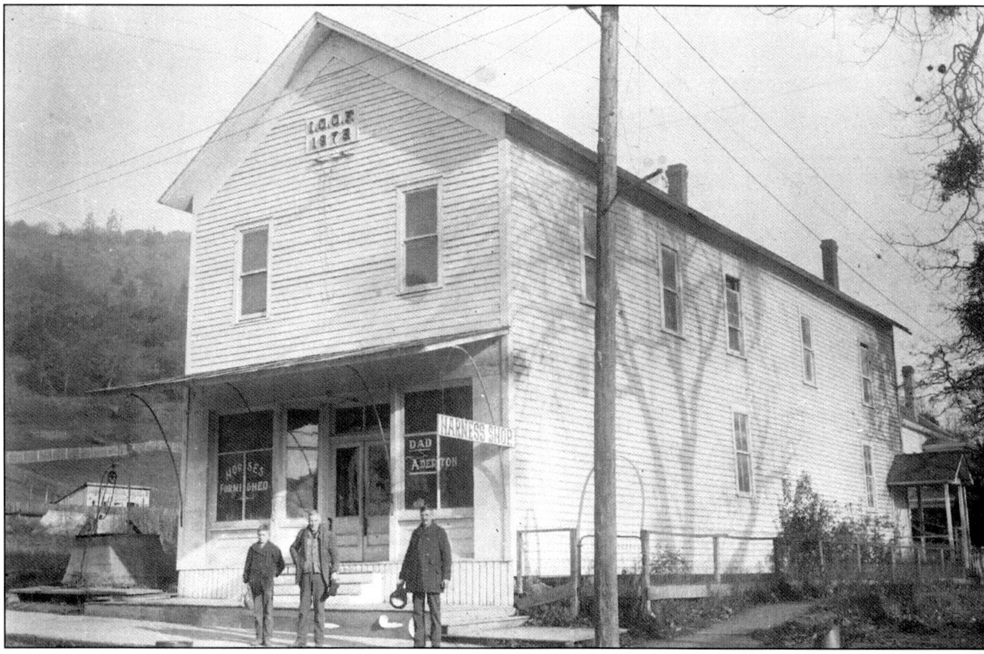

In the early part of the 20th century, "Dad" Adderton had a harness shop in this building at Third Avenue and Main Street where Shirtcliff's service station is today. Upstairs was the Independent Order of Odd Fellows (IOOF) Hall, and at the left corner of the building is the well with a pulley, rope, and water bucket. The two men and the boy are wearing heavy jackets, braced for the winter weather.

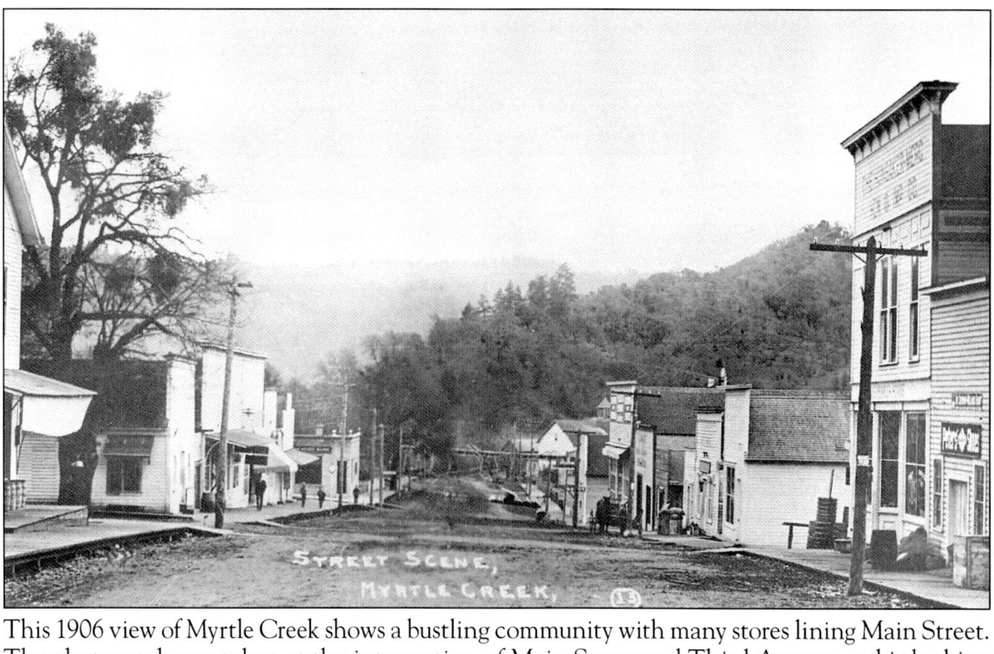

This 1906 view of Myrtle Creek shows a bustling community with many stores lining Main Street. The photograph was taken at the intersection of Main Street and Third Avenue and is looking toward First Avenue. In the right foreground is the Hunsaker Mercantile. On the same side of the street near Second Avenue is the Adams & Co. General Merchandise store.

The 1880 Myrtle Creek census shows Francis Marion Gabbert and Louisa Browning Gabbert (right) living with their 10 children. Over the years, threads of this family can be found woven into the fabric of Myrtle Creek in countless ways, and two subdivisions carry the Gabbert name. Francis Gabbert operated the gristmill, and his son Ferd owned a store and properties and was also quite the entrepreneur. Son Keeler was a druggist and a notary public, and daughter Lulu married Willis Kramer, who eventually purchased the gristmill. The Gabbert-Kramer house (below) has been a fixture in Myrtle Creek for about 110 years. Located on the northwest side of First Avenue, this multi-gabled home served for years as a boardinghouse and is now restored, still gracing the town of Myrtle Creek. (Right, courtesy of Susanna Gabbert Duke; below, courtesy of Douglas County Museum.)

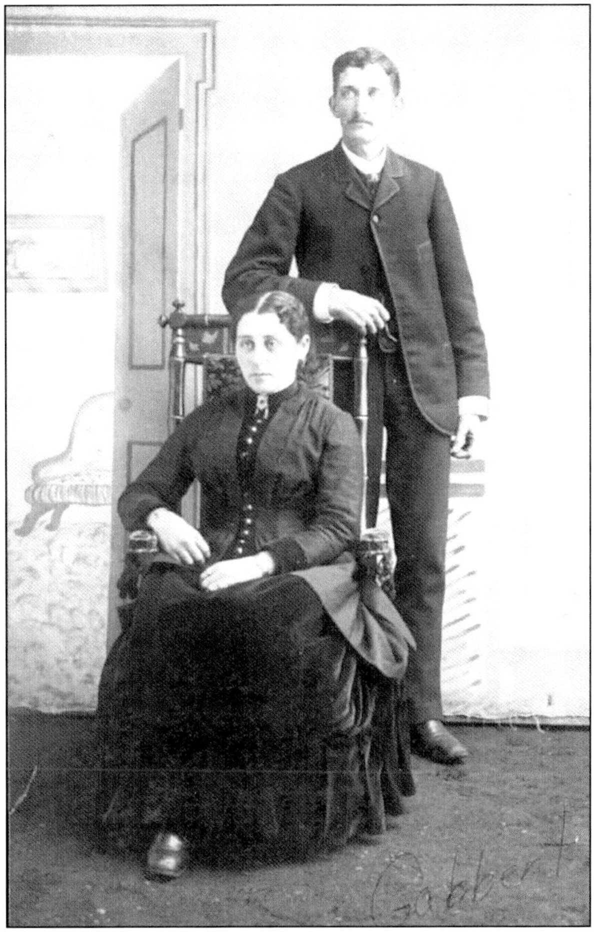

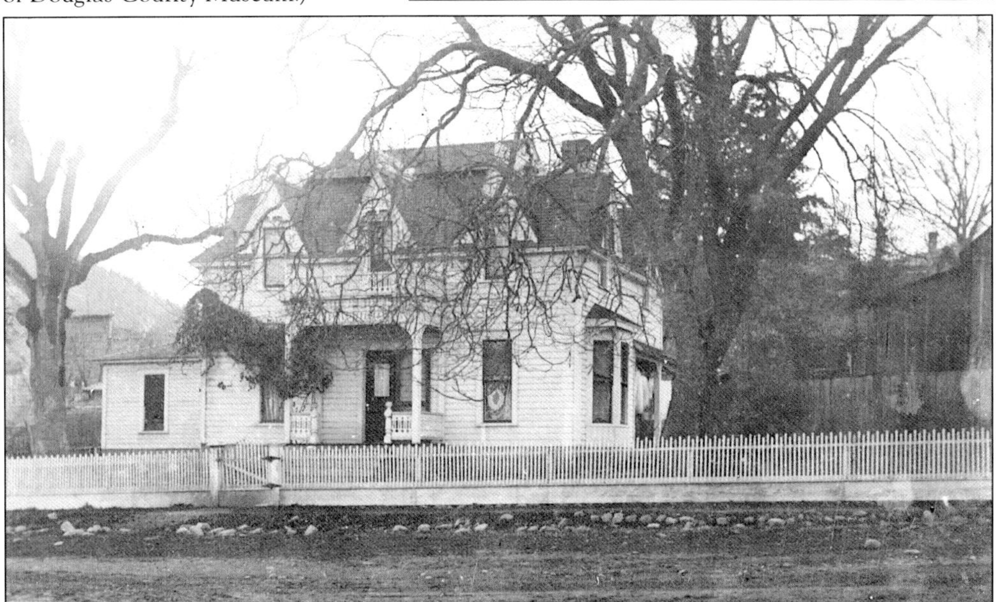

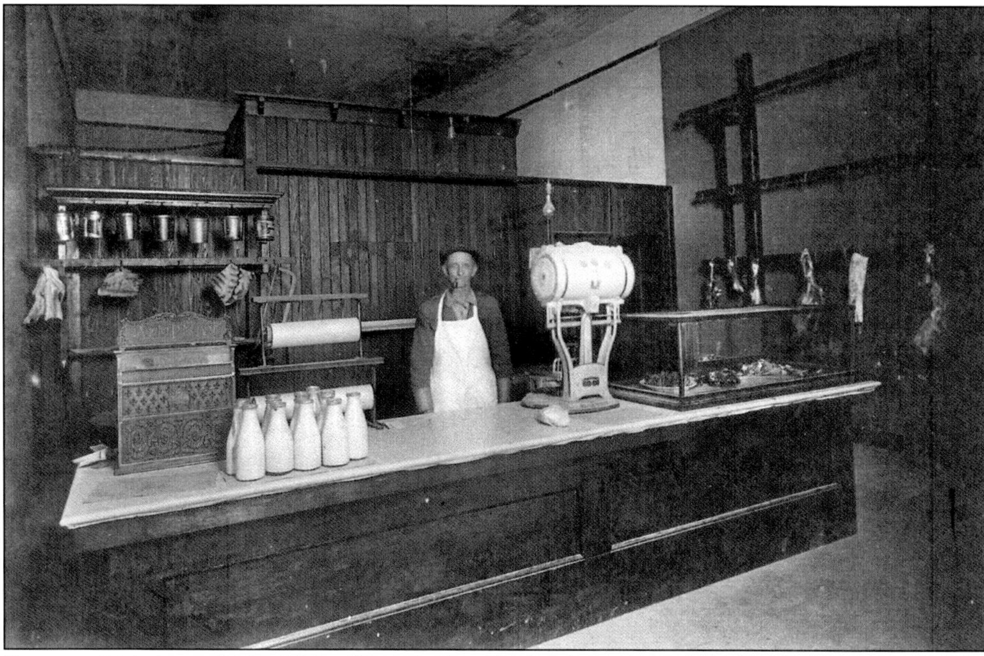

Born in 1866, Samuel "Sank" Buell worked in the local mines and on the China Ditch. He also farmed before starting his slaughterhouse and butcher shop around 1900. In this 1908 photograph, his cash register and scales share the counter with bottles of milk and a glass display case containing cuts of meat. In 1926, he sold this Second Avenue shop to the Ledgerwoods. (Courtesy of City of Myrtle Creek.)

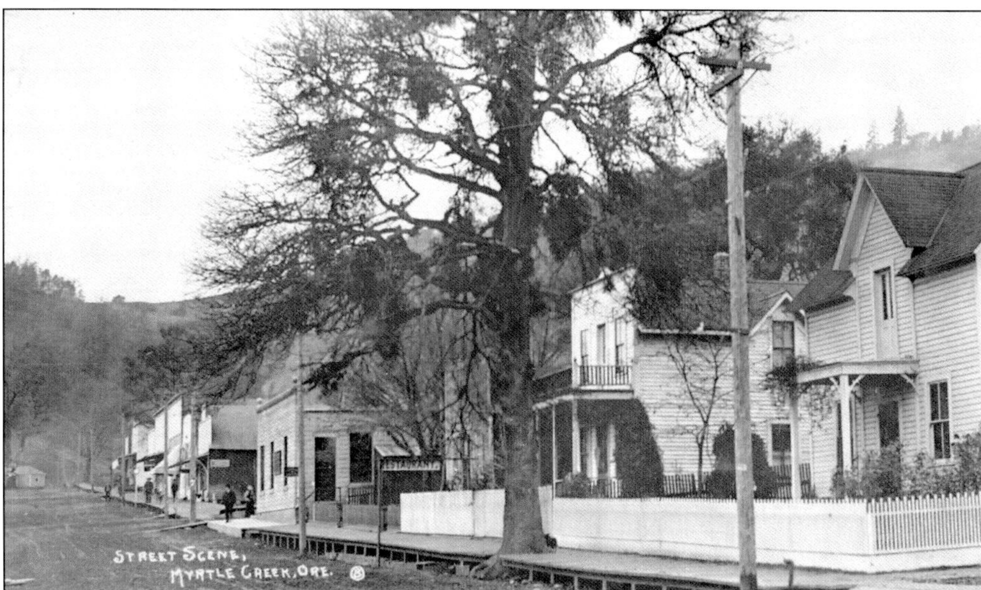

This c. 1900 photograph shows Main Street at First Avenue. The house on the right was once the home of John "Todd" Hall, son of John Hall. Years later, this house was moved to Oak Street to clear the lot for a bank. It is believed that John Hall also lived in this house around the turn of the 20th century, leaving his larger home by the gristmill to his daughter Emma Hall Buick.

Posing in front of their store on the west side of Main Street below Second Avenue are, from left to right, Harry Rice, Madge Gazley, Cina Buell, and Henry Adams. Buell, a schoolteacher, is remembered as being a spiffy dresser. Also a milliner, she made hats for the store. Later, co-owners Harry Rice and Henry Adams partnered with James Rice and in 1915 built the Rice Bros. & Adams General Merchandise, an expanded store and warehouse.

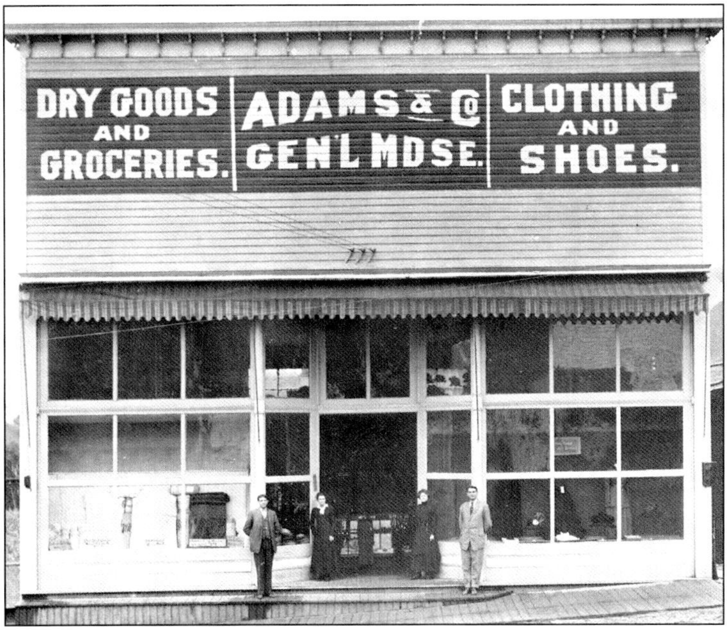

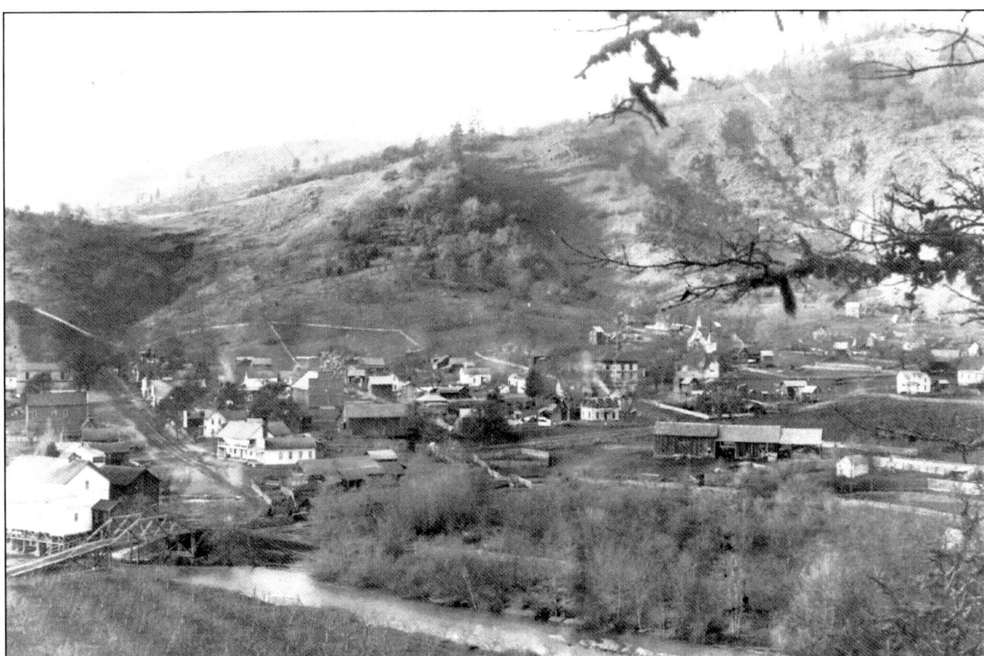

This panorama of Myrtle Creek was taken before 1903 and shows the wood truss bridge next to the gristmill. Hall's prune orchard can be seen in the left foreground. The angled building across from the mill is the Overland Hotel, built in 1890; it burned down in 1905. Also visible are the school, the Christian and Presbyterian Churches, and the Gabbert house.

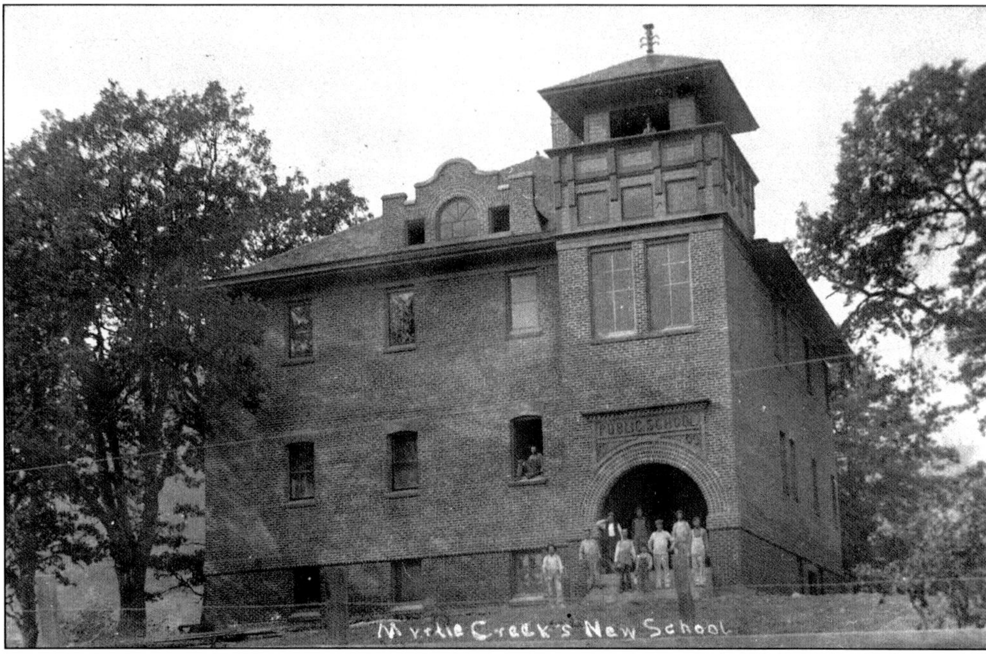

In 1909, construction workers pose in front of the new brick school being built on Division Street using bricks quarried and fired on the site. In 1923, a wing added to the right side provided additional high school classrooms. On Easter Sunday 1929, a fire destroyed this building, and the current building (now the grade school, shown below) was built to replace it.

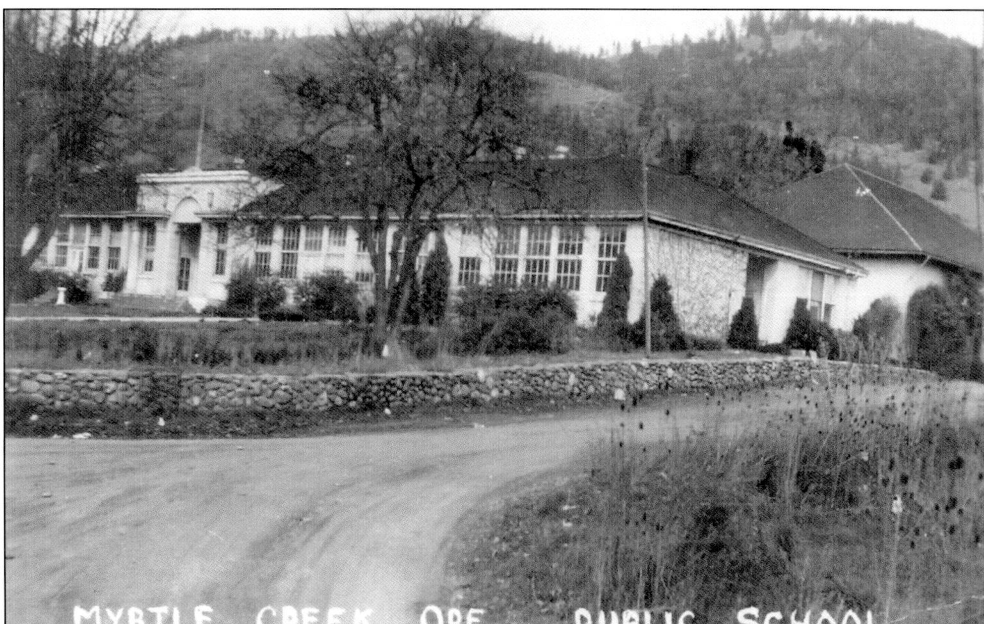

Built as a replacement for the school lost in the 1929 fire, this building was converted to an elementary school in 1965 when South Umpqua High School was built in Tri City. Eighty years after it was built, it is still in use as a school and still graced by beautiful, natural river rock retaining walls.

Three

100 Valleys of the Umpqua

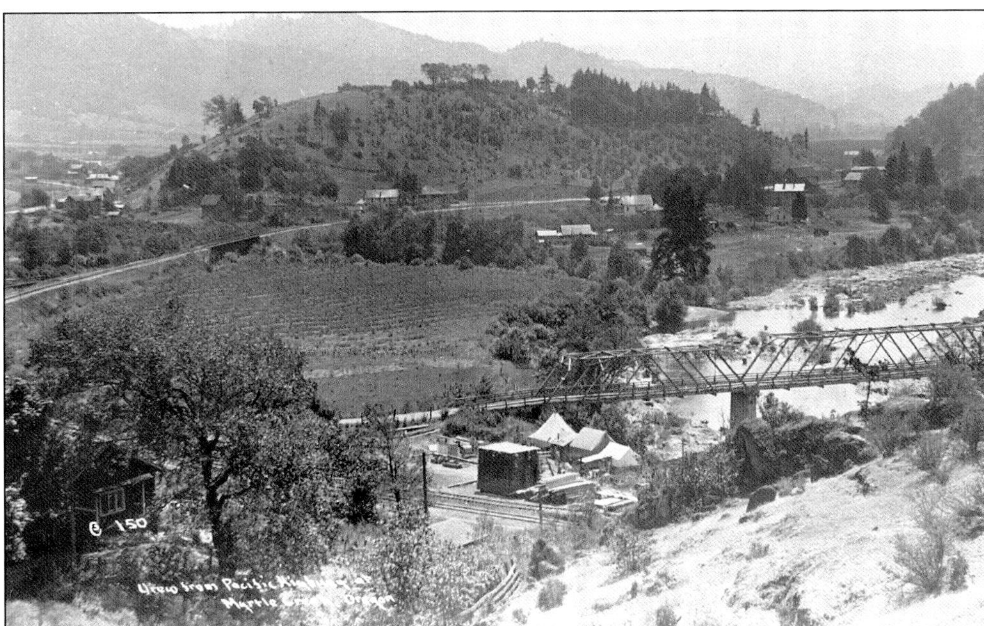

Pictured here is a new, third way into town. Of limited duration, this bridge served the residents of Riddle, Weaver Road, and the Boomer Hill area, but it was not the main thoroughfare. Unfortunately, little is known of this bridge that connected to Fourth Avenue. The Oregon Department of Transportation has no record of it, so it is not known when it was built. Photographs show it being dismantled after the Pacific Highway 99 bridge was completed in 1922.

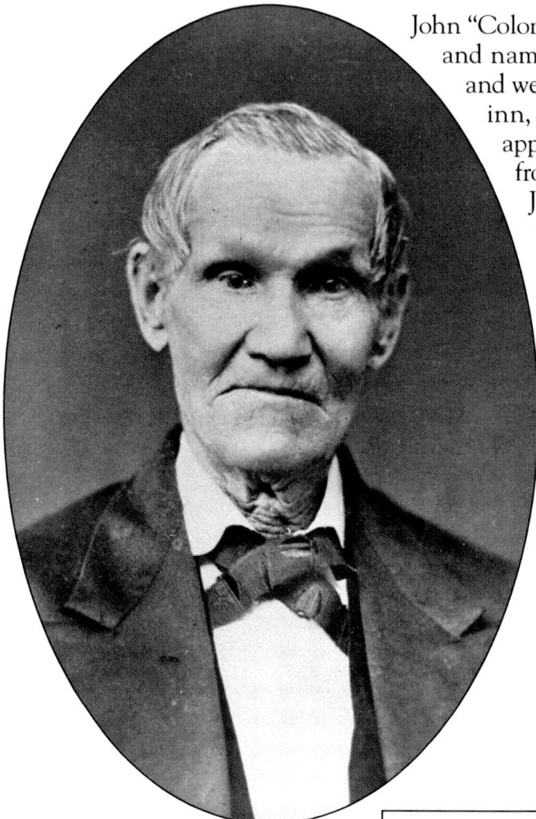

John "Colonel" Sims Burnett, a family patriarch, settled and named Round Prairie in 1852. Civically active and well regarded, his home served as a post office, inn, and community center. In 1852, Congress appropriated $120,000 to build a military road from Scottsburg to California and assigned Capt. Joseph Hooker (Gen. Joe Hooker of Civil War fame) to construct the road. Captain Hooker was a frequent guest of Burnett during the construction.

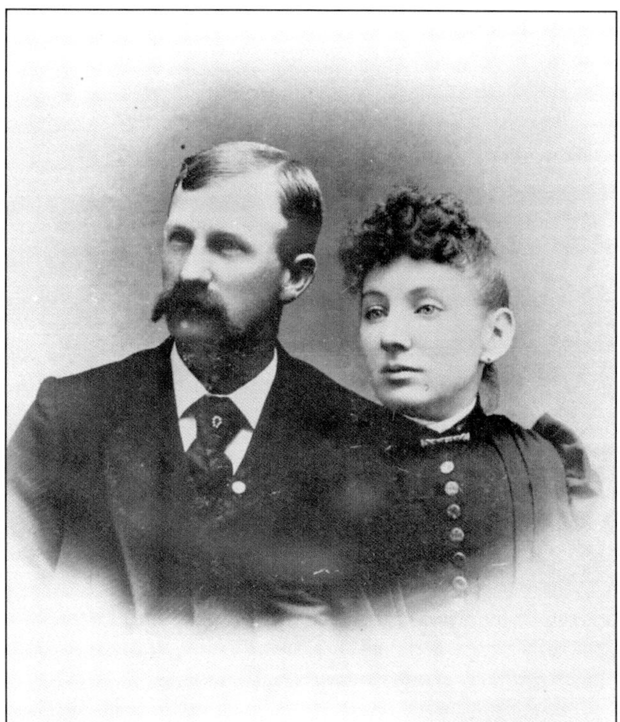

John Burnett's grandson Tom, pictured with wife Gloria, began his career at age 17 by driving a four-horse stagecoach from Round Prairie to Roseburg. He drove the local stagecoaches until the railroad came south from Roseburg in 1882, and then he drove in northern California. After retirement, he returned home and often drove his stage in parades. Tom's stage is now housed in the Douglas County Museum.

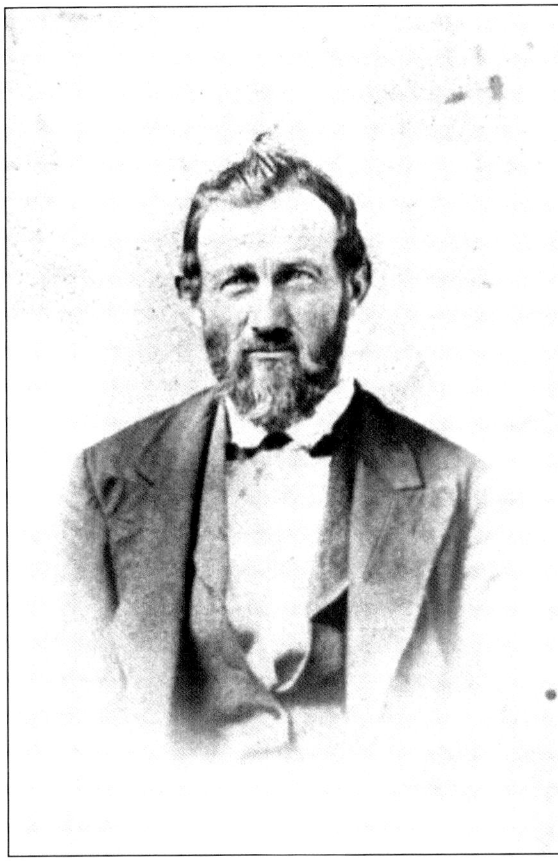

James Burnett (right), son of John Burnett, settled with his father and brother on separate Donation Land Claims in Round Prairie. James's house, built in 1856, is pictured below shortly before it was razed in 1916. James was active in civic affairs and served as a Douglas County commissioner. He was the father of Tom Burnett, the noted stage driver pictured opposite.

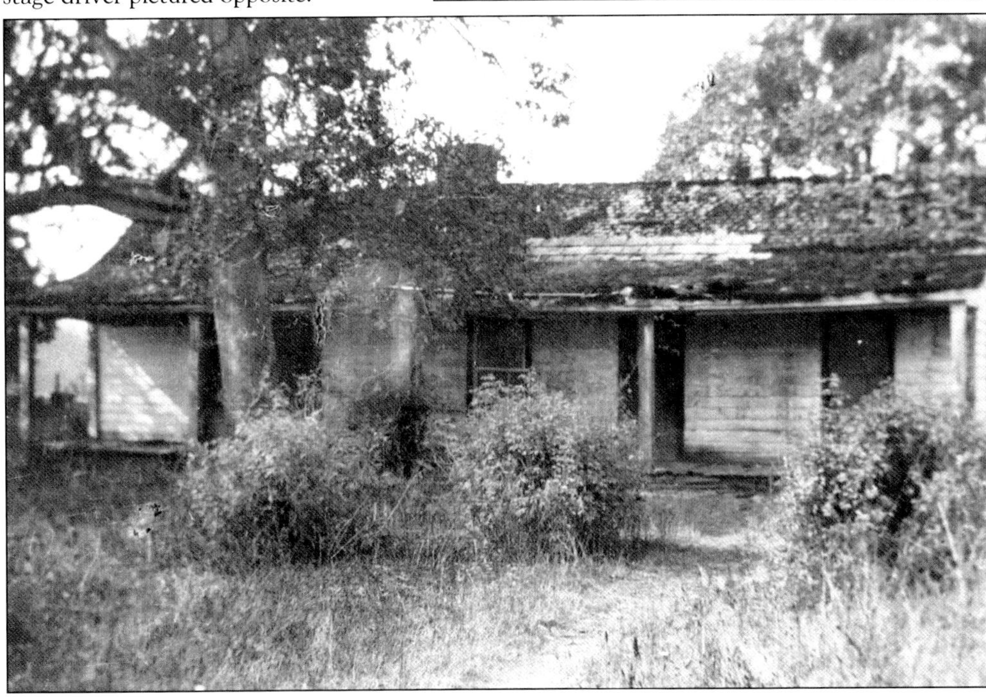

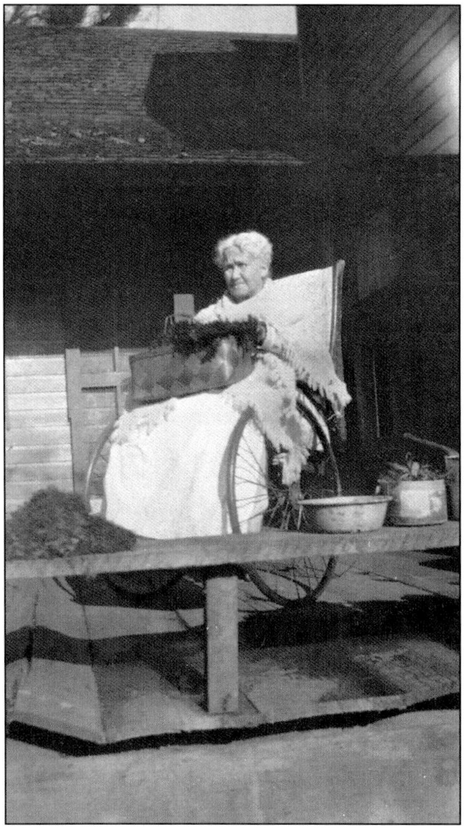

Mary Roberts, daughter of Jesse Roberts (of Roberts Creek and Roberts Mountain), married George Stevenson of Oak Grove, later called Ruckles or Dole. They operated a popular stage house and inn while farming the flat along the river and also logged the hills to the east. The Douglas County Museum has a wonderful account that Mary left of events and life in the 1850s.

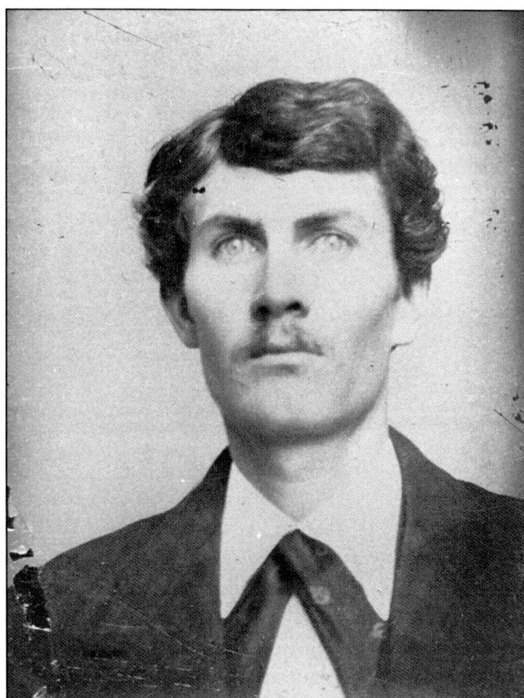

Joseph Lane Jr. (1827–1907) was the son of Gen. Joseph Lane, the territorial governor of Oregon. He fought beside his father in the Mexican War and then came to Oregon in 1851. In 1854, he married Eleanor, daughter of Samuel Stevenson of Oak Grove. With their eight children, they moved to the Myrtle Creek area, where he served as justice of the peace until his death.

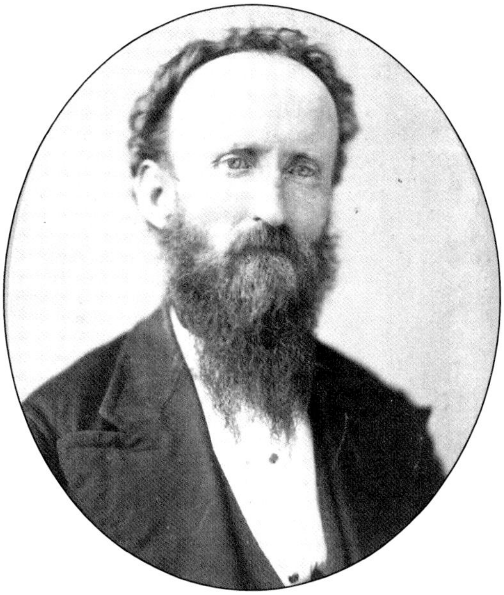 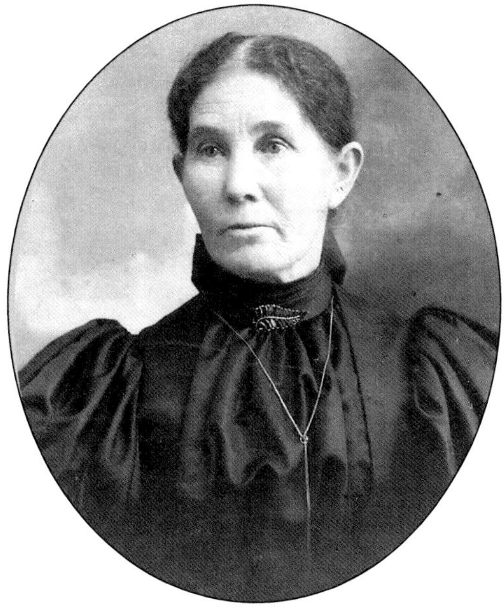

Bernard Pitzer Smith (1823–1882) married Susan Dickenson (1836–1905) in 1855. In 1866, they bought the Michael Hanley place, located about five miles north of Myrtle Creek. In 1882, Bernard was killed at Oak Grove in a wagon accident caused by a train scaring the horses. The ranch then went to George Pitts, then James Booth, and finally to the Helgeson family. (Courtesy of Sandy Smith Hilburn and Al Helgeson.)

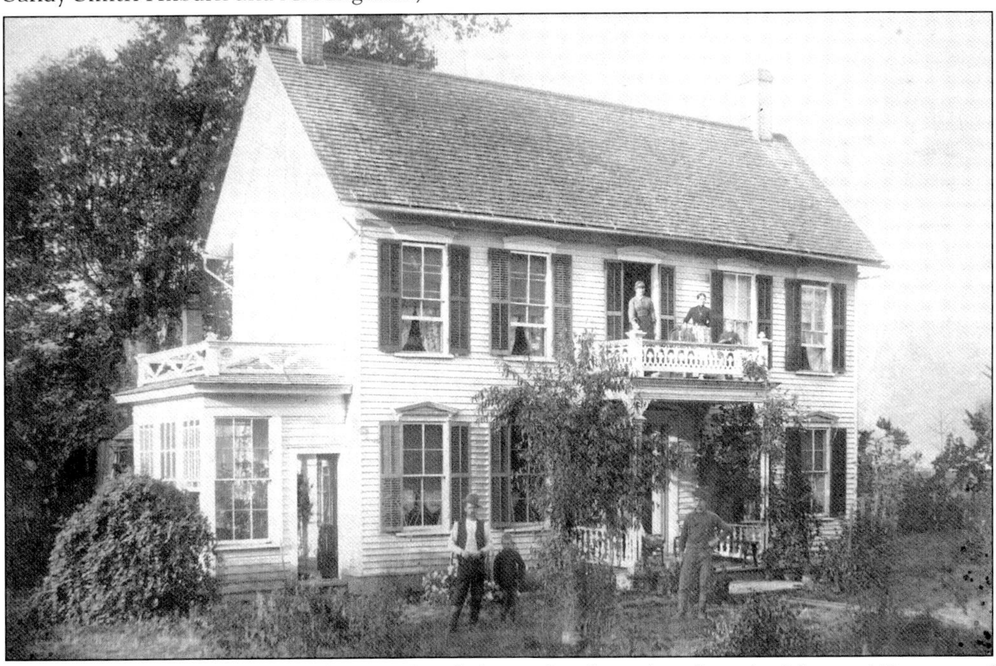

Built around 1870, the Bernard Pitzer Smith house has been listed in the National Register of Historic Places since 1989. Located on Old Highway 99 about six miles north of Myrtle Creek, it was rotated in 1923 by the Booth family to face the newly built Pacific Highway. The Helgeson family purchased this beautiful house in 1943; 69 years later, it is still occupied by a son, Alvin, and his wife, Jackie Helgeson. (Courtesy of Al Helgeson.)

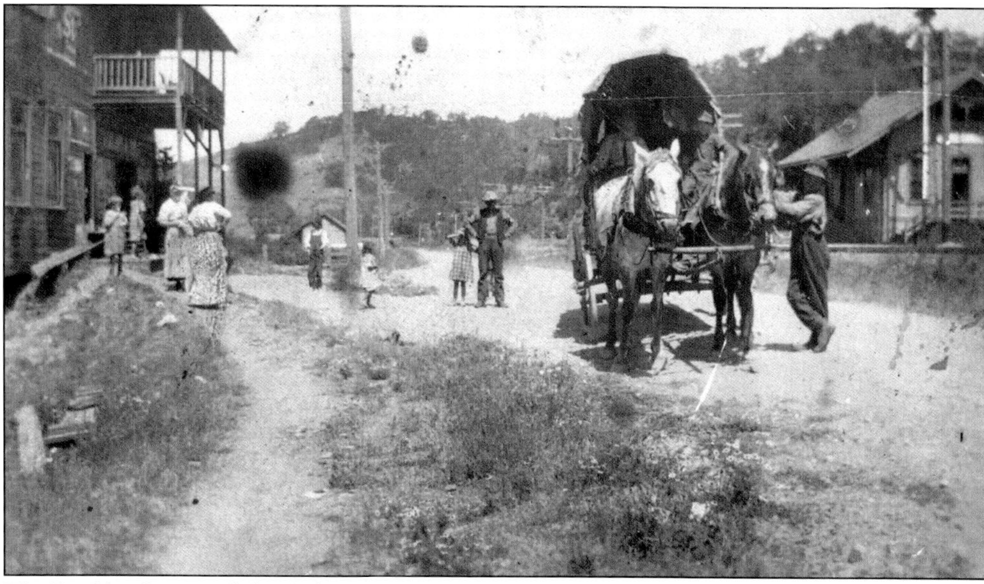

The Ruckles Store and the Dole station are pictured around 1923. Originally called Oak Grove, the name was changed to Ruckles when the name "Oak Grove" was denied by the post office because of name duplication. A traveler's inn started here by Samuel Stevenson in the late 1850s became a stage stop, which was taken over by Matt Ruckles in 1864. (Courtesy of Gladys Russell Schartner and Dorothy Shartner White.)

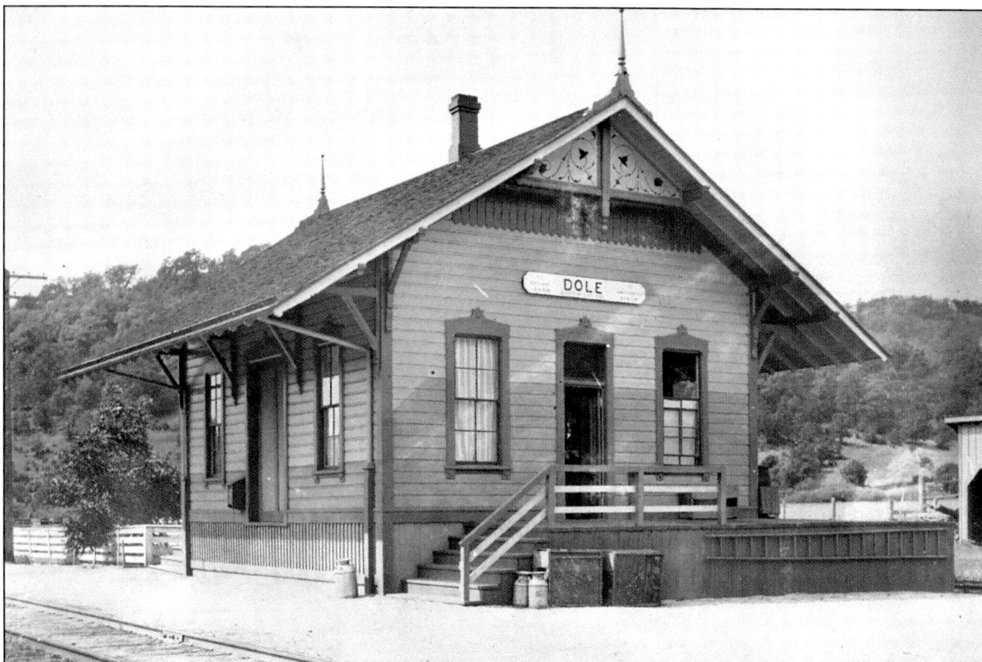

Located 555 rail miles from San Francisco and at an elevation of 597 feet, the Dole railroad station and siding was located four miles north of Myrtle Creek, but only the siding remains today. Originally called Ruckles, the name was changed to Dole by the O&C Railroad to avoid confusion with Riddle's Station.

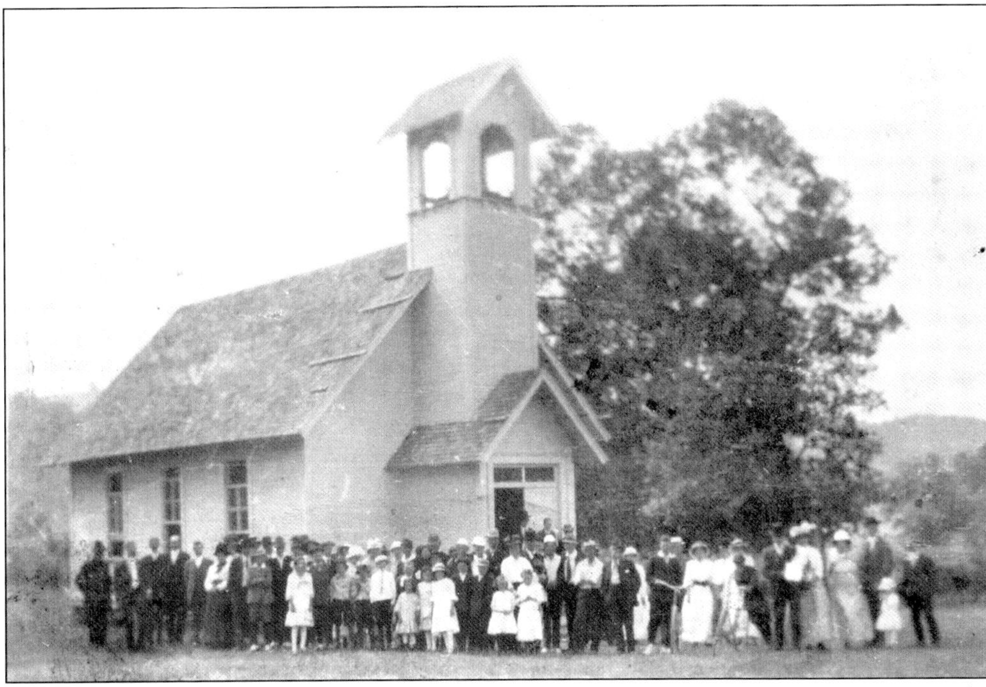

The Ruckles Church was located in the vicinity of what is now called Clarks Branch. As this c. 1920 photograph shows, it was well attended. Though people in these small communities were closely associated with their church, the railroad provided another alternative. On scheduled days, a Chapel Car arrived in town complete with an organ, pulpit, and preacher to minister to the needs of the local population. (Courtesy of Al Helgeson.)

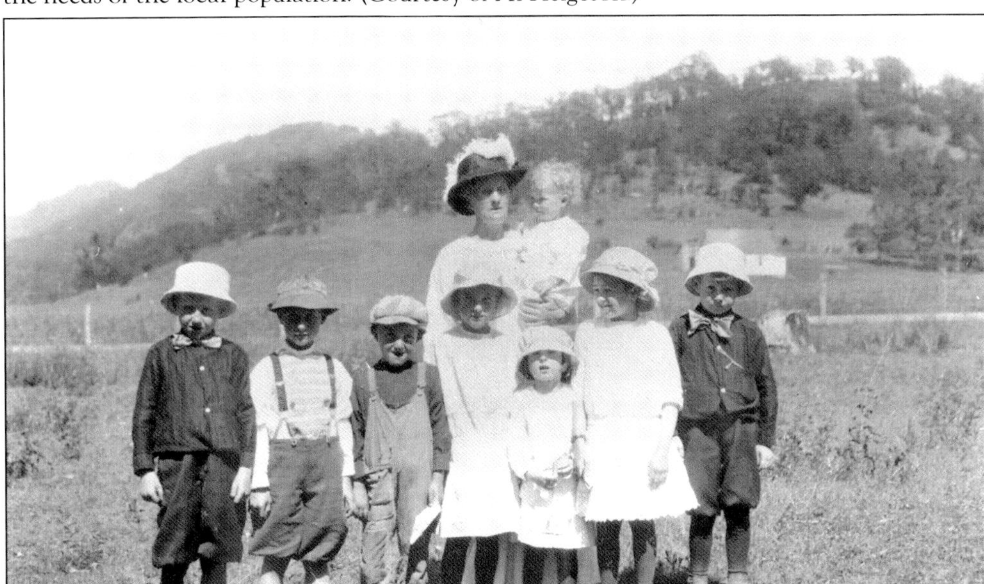

The Ruckles School, perched on the hill behind these students and their teacher, was attended in the 1880s by a student named Clarence Eddy. He later gained fame as the "Poet-Prospector." Not only famous for the productive mines he discovered in five western states, he wrote several published books of poetry. One of them, *Pinnacle of Parnassus*, can still be found online today.

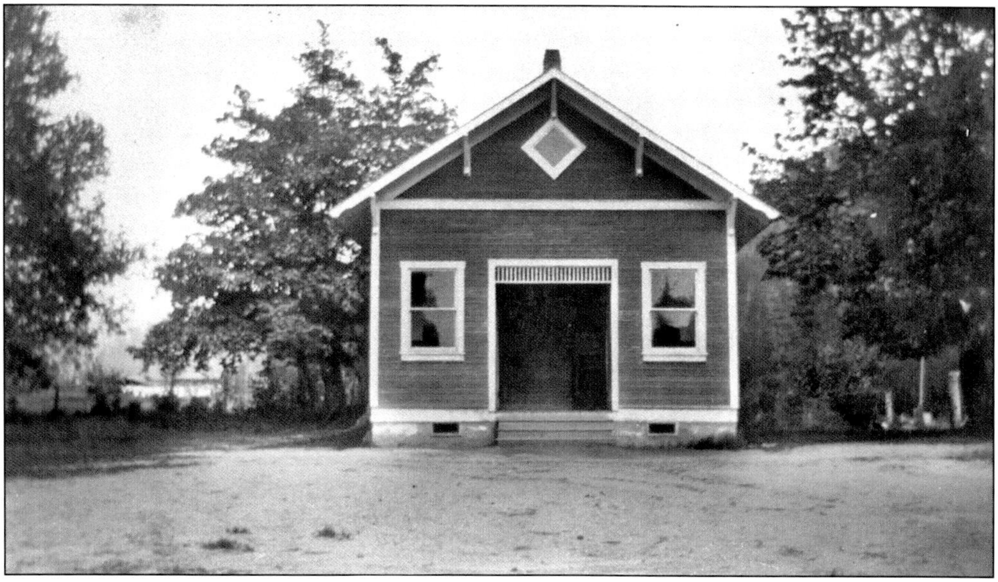

Neater and cuter than the typical rural schoolhouse, the prosperous mining community of Nugget took pride in their nicely painted school. Located about 11 miles up South Myrtle Road, several current local citizens began their education in the Nugget school. In the 1930s, the school sponsored a successful carnival, and the $6.75 in proceeds were used to provide hot soup for the students on cold winter days.

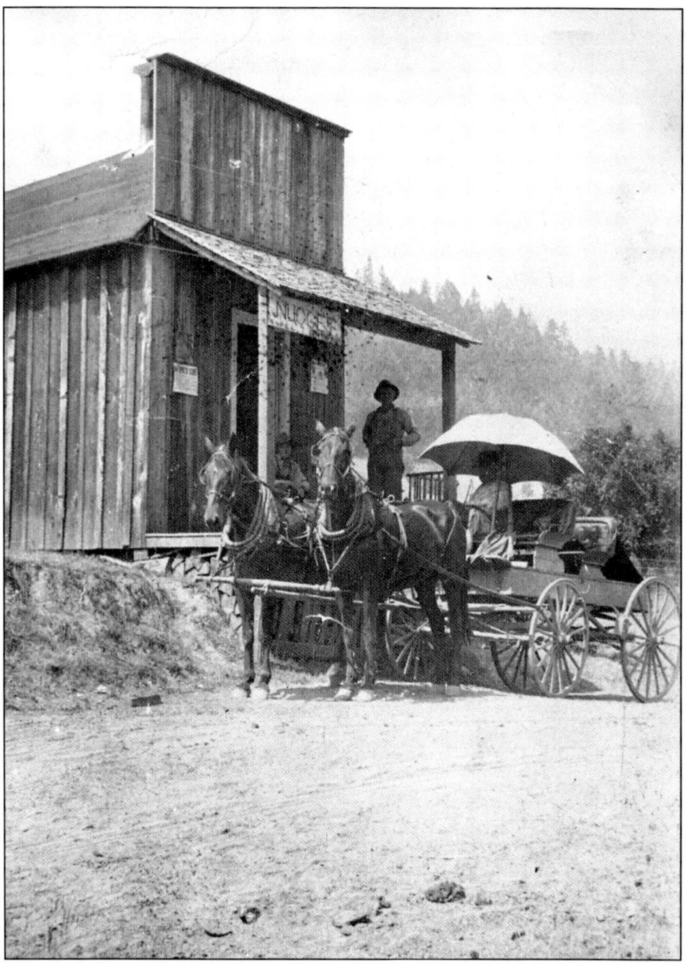

Parked in front of the Nugget Post Office around 1920 is Ethel Boyle, the mail woman, in her horse-drawn hack. She is protected from the elements by a big "boomer." In those days, the mail carrier also served as a delivery and taxi service for people living along the route.

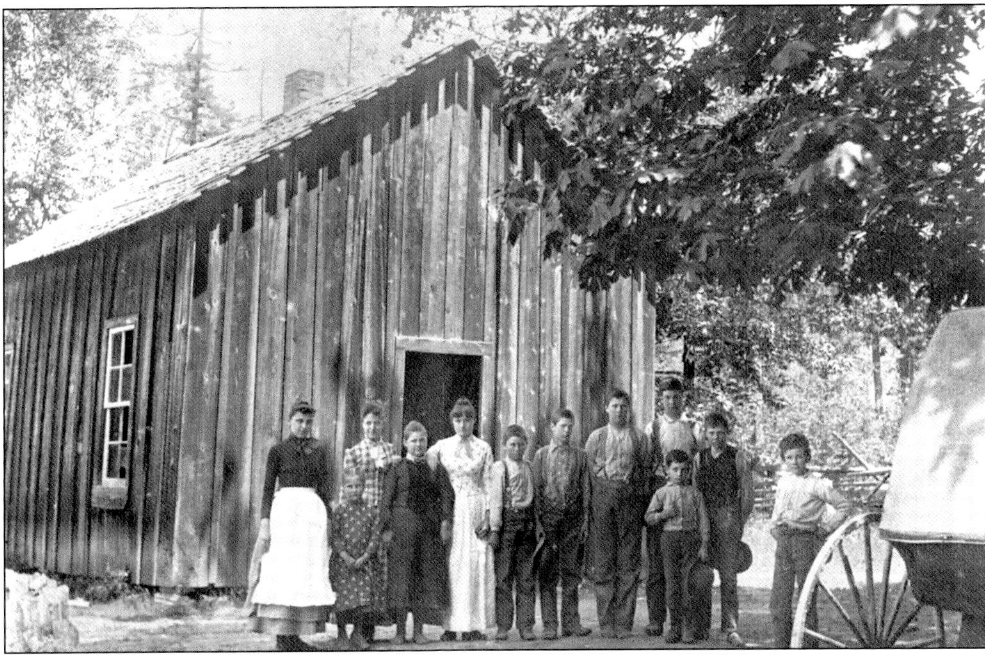

The Dyer School, located near the mouth of Strode Canyon five miles up South Myrtle Creek, was in use until 1919, when it incorporated with the Myrtle Creek School District. Below is a damaged—but rare—photograph of the interior of the school. The cameras of the time required too much light to allow for many interior photographs, so it is a nice treat to see the inside of a classroom in an old rural school. This class of diligent students includes Art Dyer, who is seated in front of the window in the back to the right. (Above, courtesy of the Douglas County Museum; below, courtesy of Gladys Shelton Grelck.)

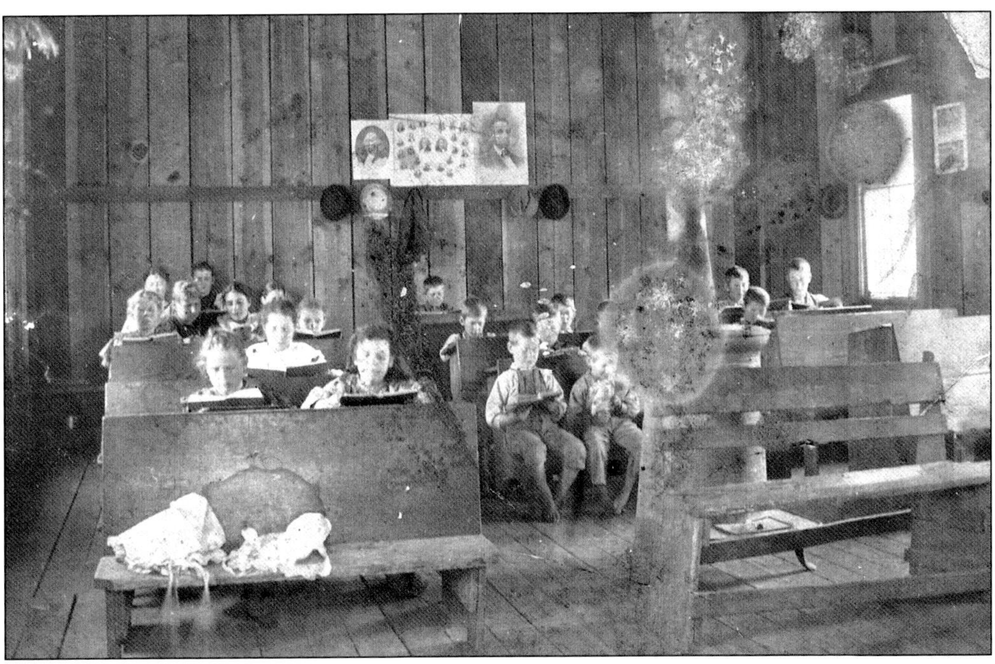

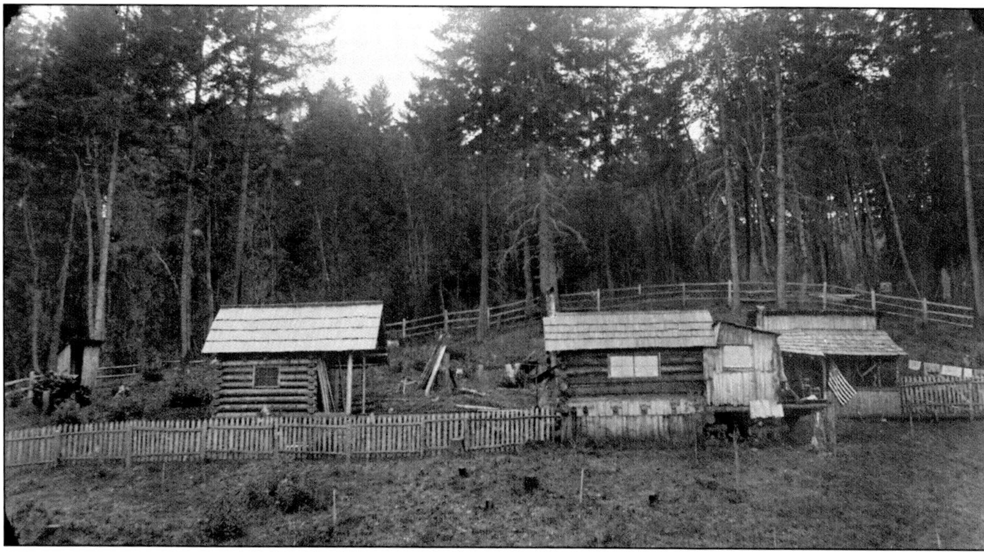

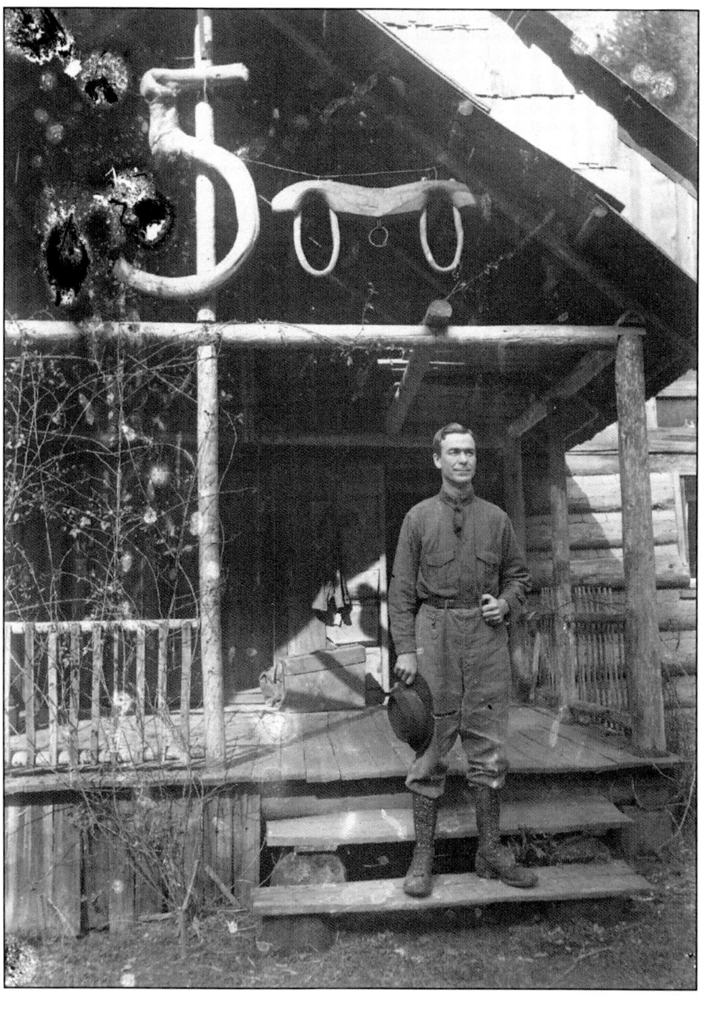

This early 1900s photograph of Buck Fork, about 13 miles up North Myrtle, shows the flag flying from the post office. Buck Fork had postal service from 1910 to 1927. The photograph on the left shows another Buck Fork Post Office site in the home of an unnamed man. Primarily a logging community, the town housed families until the late 1940s or early 1950s.

The Depression-era photograph at right shows a group of barefoot Buck Fork children clad in clothing that was probably patched hand-me-downs. The first Buck Fork School was originally a chicken house, but it was replaced in 1925 by a new building. Below, the Buck Fork School, perched on a hill above North Myrtle Creek, is shown with students from the Dunnavin, Thackery, Conley, and Moore families.

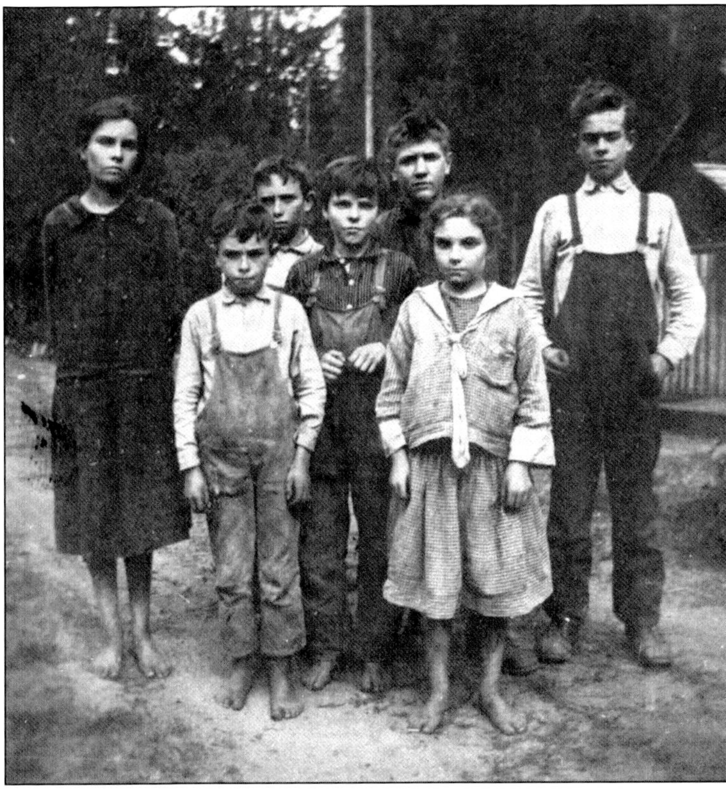

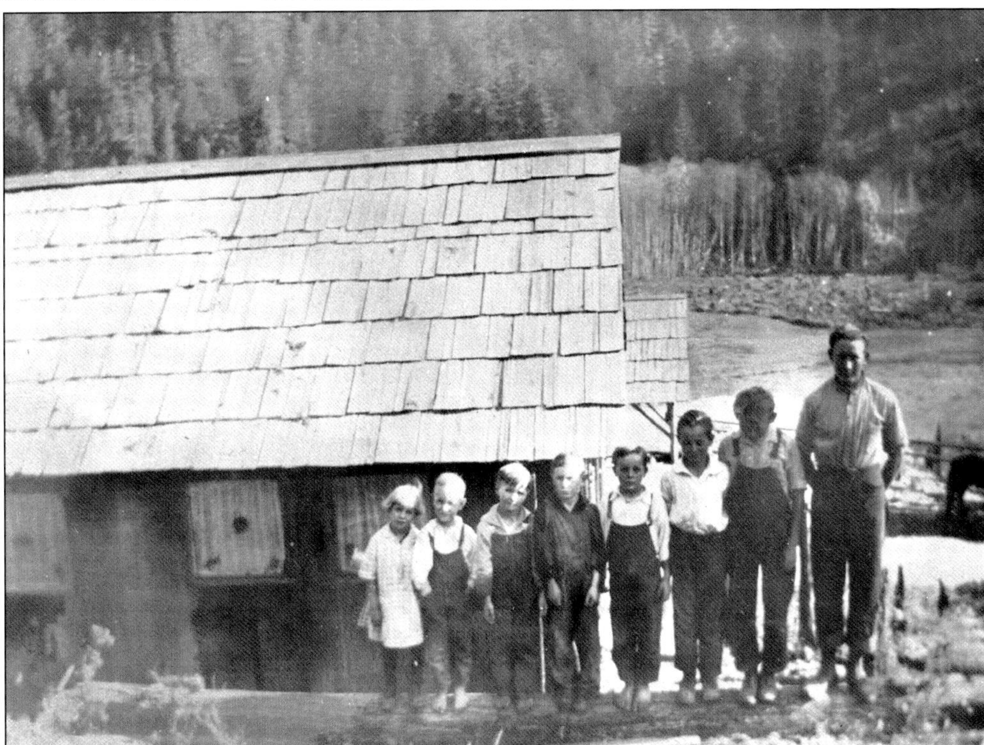

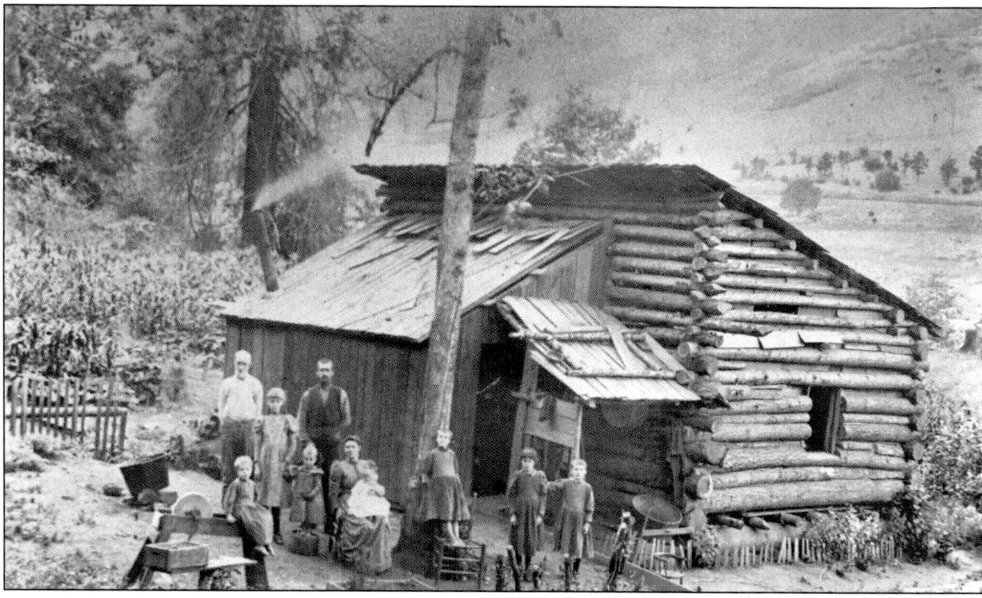

The Findlay place (above), just north of Myrtle Creek on the west side of the river, and the Worthington place (below) are examples of late 19th and early 20th century rural living. Plentiful are the photographs of fancy homes of urban citizens, but few are the photographs of the hardscrabble homes of rural citizens.

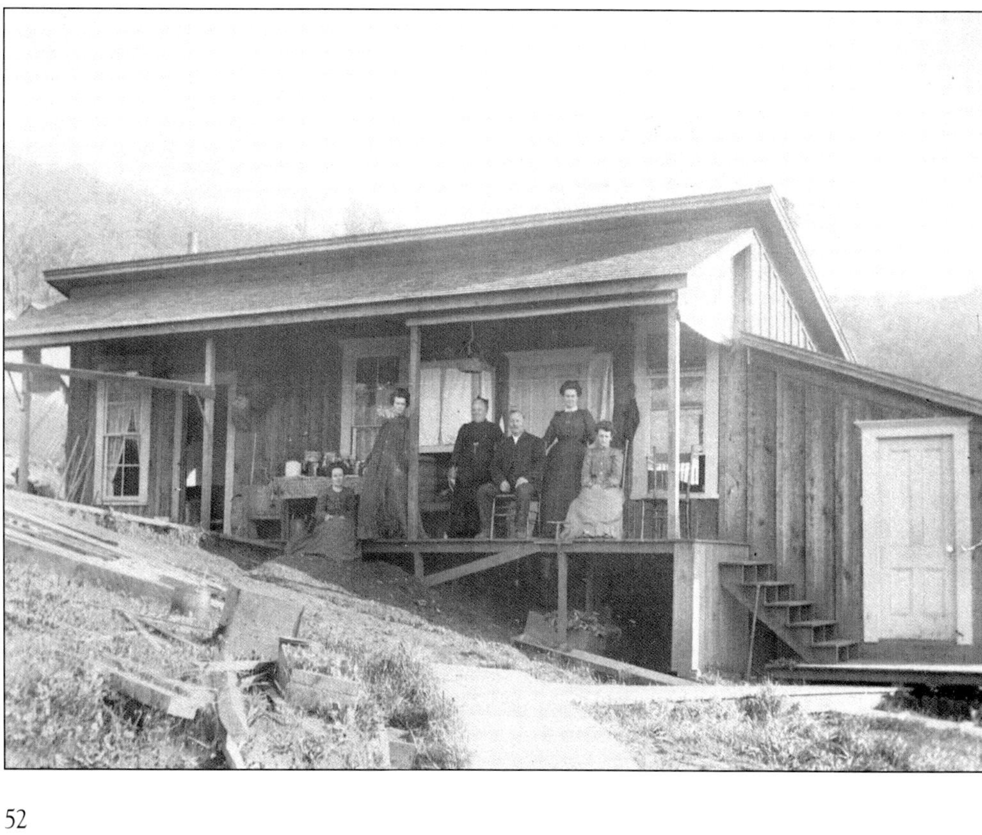

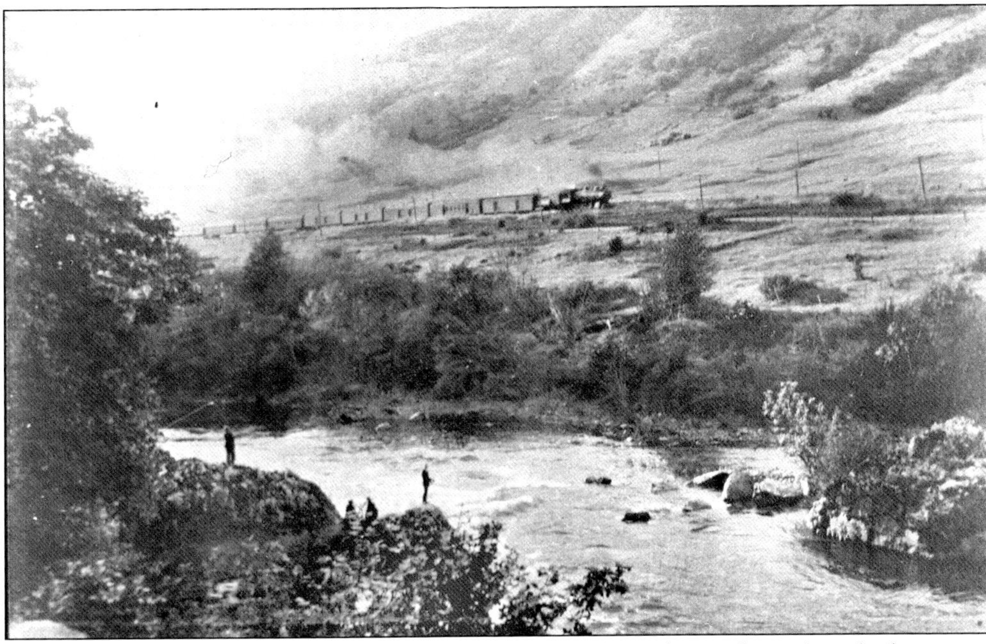

As the southbound train crosses Hadley Flat just north of Myrtle Creek, anglers ply their art in the foreground. Quoted in a 1935 newspaper, local angler Tom Cornutt claimed that at times the salmon were so thick in Myrtle Creek that they smothered all the little fish. Contemporary anglers, still fishing from the same rock, can no longer make that claim.

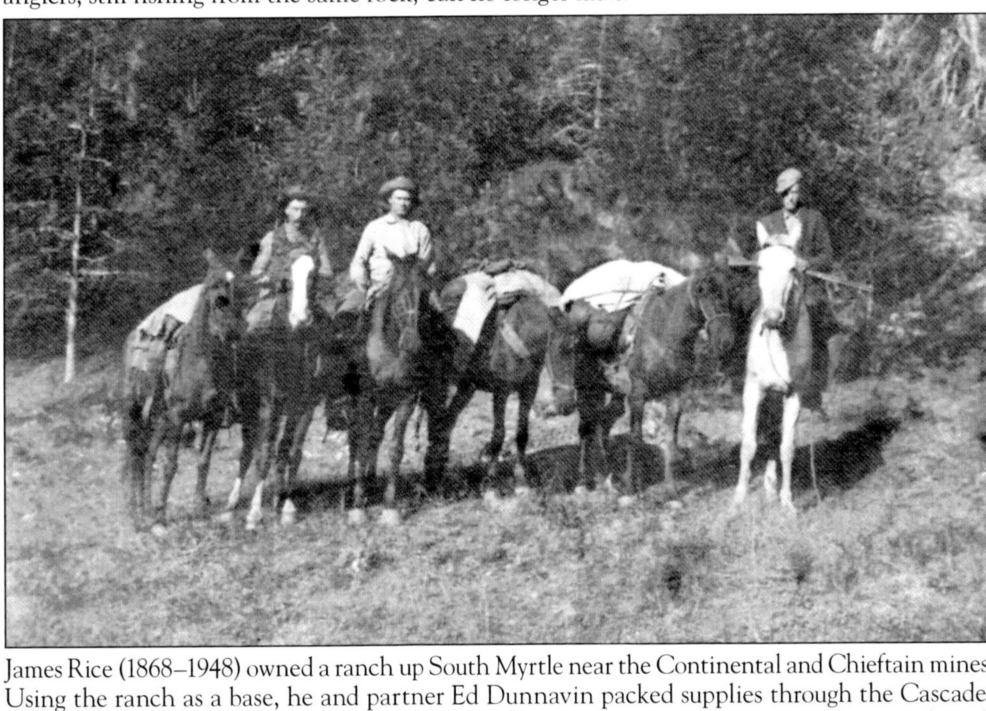

James Rice (1868–1948) owned a ranch up South Myrtle near the Continental and Chieftain mines. Using the ranch as a base, he and partner Ed Dunnavin packed supplies through the Cascades to a ranch they shared at Warner Valley in southeastern Oregon. This c. 1900 photograph with Dunnavin (left) and Rice packing a rifle (right) was taken near the head of South Myrtle Creek. (Courtesy of Bill Rice.)

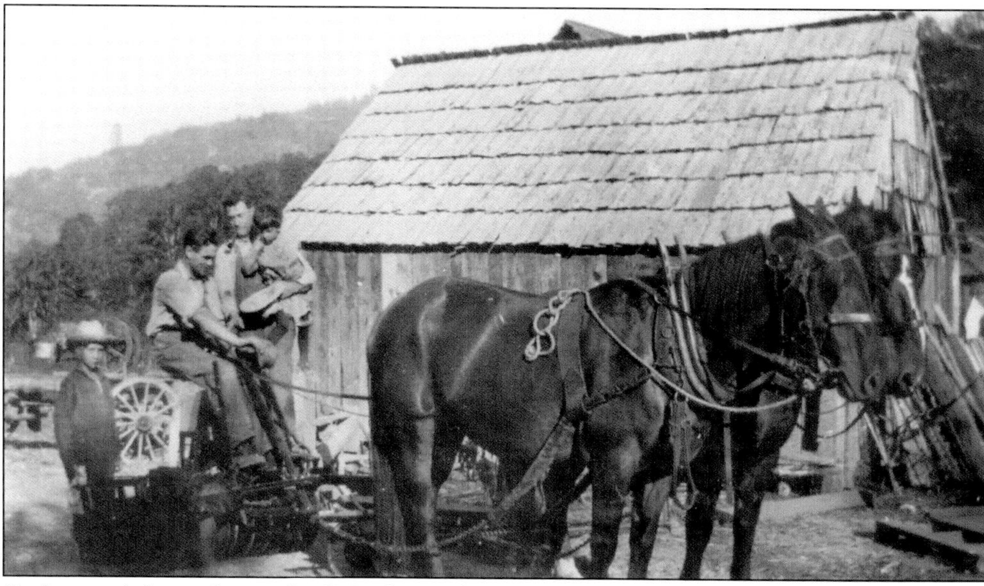

On his ranch five miles north of Myrtle Creek, Ora Ronk and his sons pose with a team of horses hitched either to a disk or "cultipacker." After a field was plowed, a disk broke up the clods. Then, depending upon soil condition and the type of crop to be planted, a cultipacker might be used to lightly pack and condition the soil for planting. (Courtesy of Al Helgeson.)

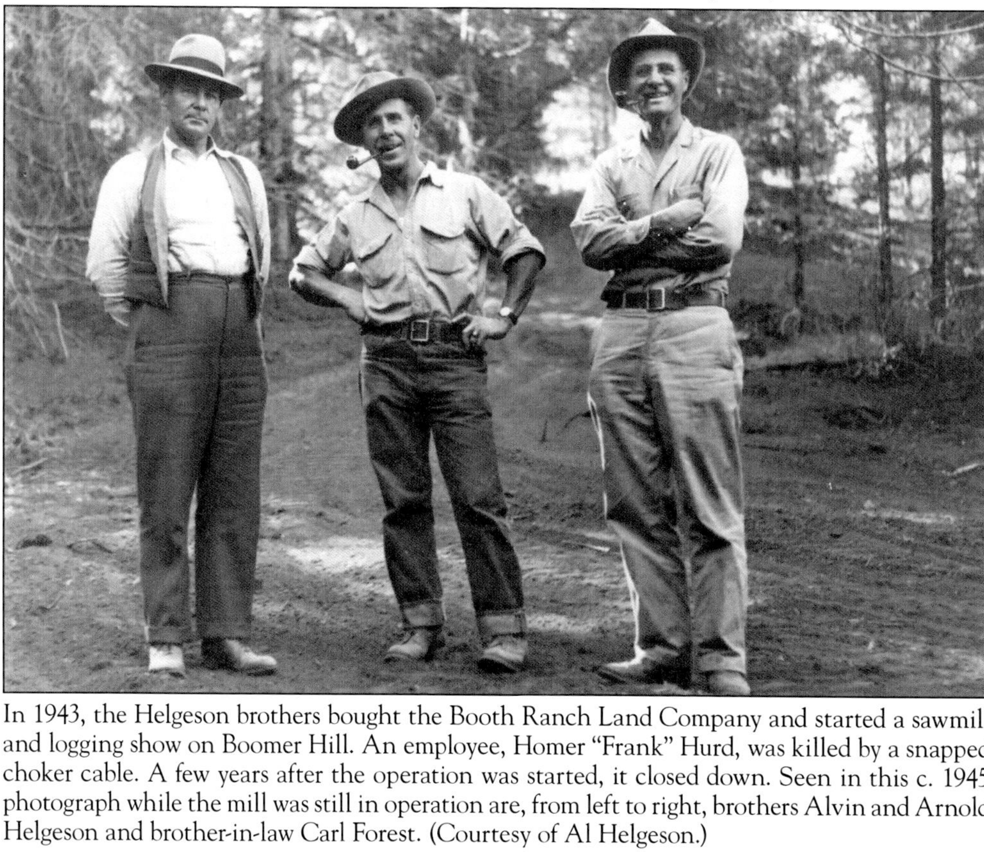

In 1943, the Helgeson brothers bought the Booth Ranch Land Company and started a sawmill and logging show on Boomer Hill. An employee, Homer "Frank" Hurd, was killed by a snapped choker cable. A few years after the operation was started, it closed down. Seen in this c. 1945 photograph while the mill was still in operation are, from left to right, brothers Alvin and Arnold Helgeson and brother-in-law Carl Forest. (Courtesy of Al Helgeson.)

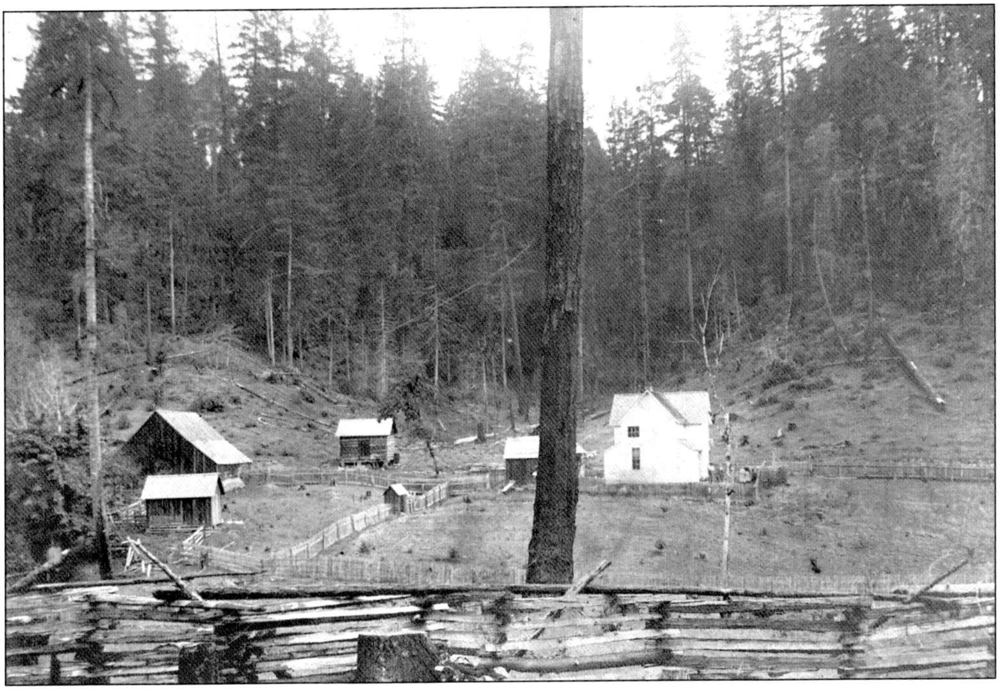

The James Andrew Jackson property on Frozen Creek (above) and the John Bowman place on South Myrtle Creek (below) are representative of many rural landowners of the time. These c. 1900 photographs show cleared land, fences, garden plots, outbuildings, and well-tended homes, all signs of the amount of time and labor that were invested into the making of a home.

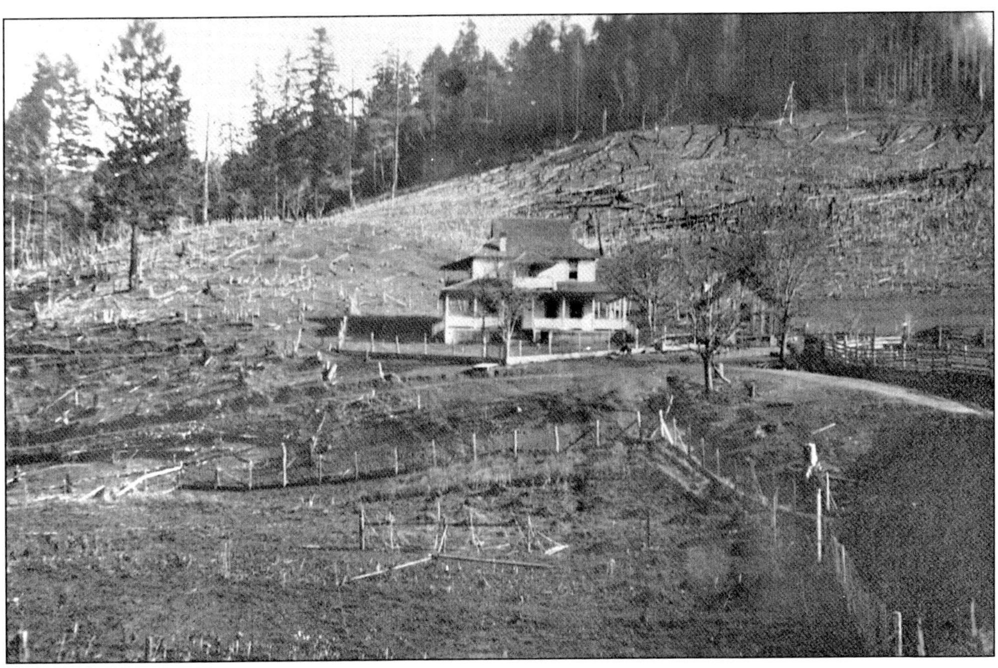

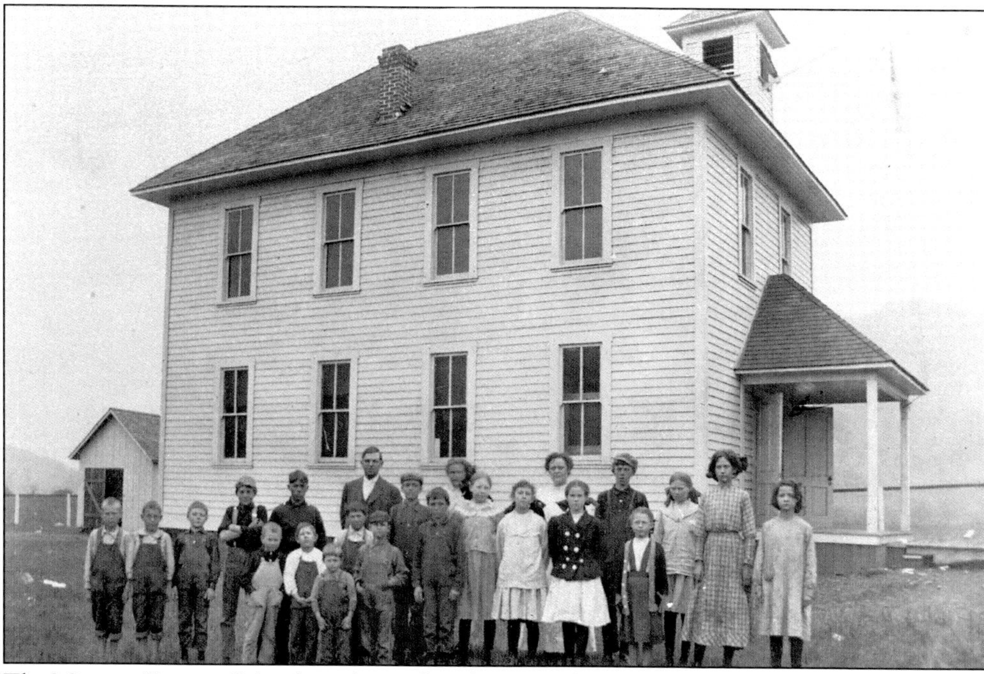

The Missouri Bottom School was located in the general vicinity of South Umpqua High School in Tri City. Names for the students in this 1911 photograph are given, but there are more students than names, making identification difficult. The teacher is Alvin Tipton, and he was instructing children from the Flynn, Jones, Foltz, Weaver, Cornutt, and Arzner families.

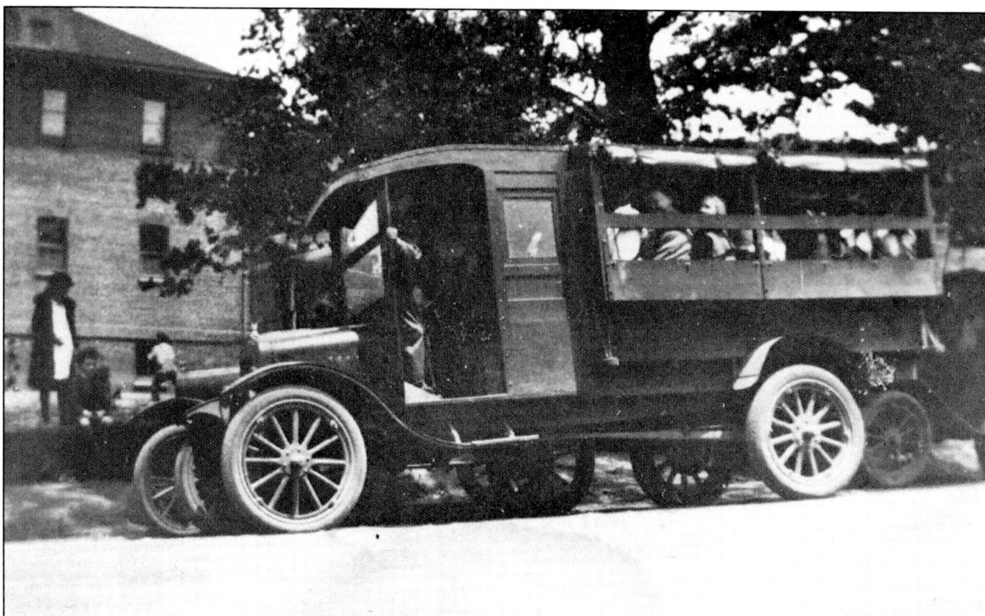

This 1924 Model TT Ford (second "T" for truck), outfitted with roll-up canvas curtains, served as a school bus for both North and South Myrtle students. It must have been a durable, memorable bus as it survived several drivers, hundreds of kids, and is frequently mentioned in memoirs found in local historian Blanche Newton's *The Golden Leafs of Time*.

Four

Prunes and Prosperity

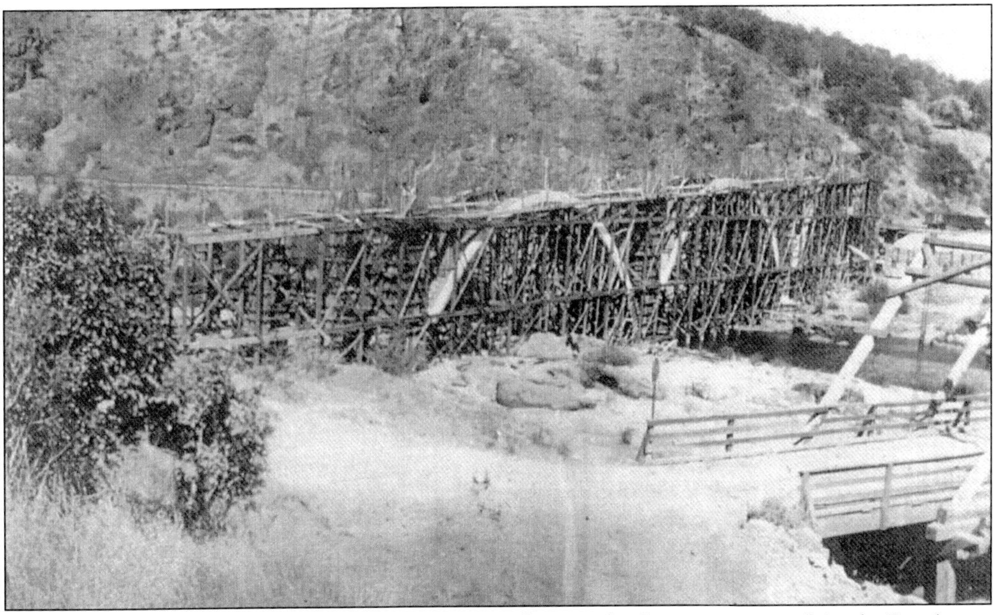

With the construction of Pacific Highway 99 and a new bridge in progress, Myrtle Creek was about to welcome the onset of the automobile era. Served for 70 years by the original Scott-Applegate Trail with improvements for wagon use, the new highway was the first route to town built specifically for automobiles. In 1922, Main Street was rerouted to connect to the new bridge. (Courtesy of Sharon Lozano and Blanche Newton.)

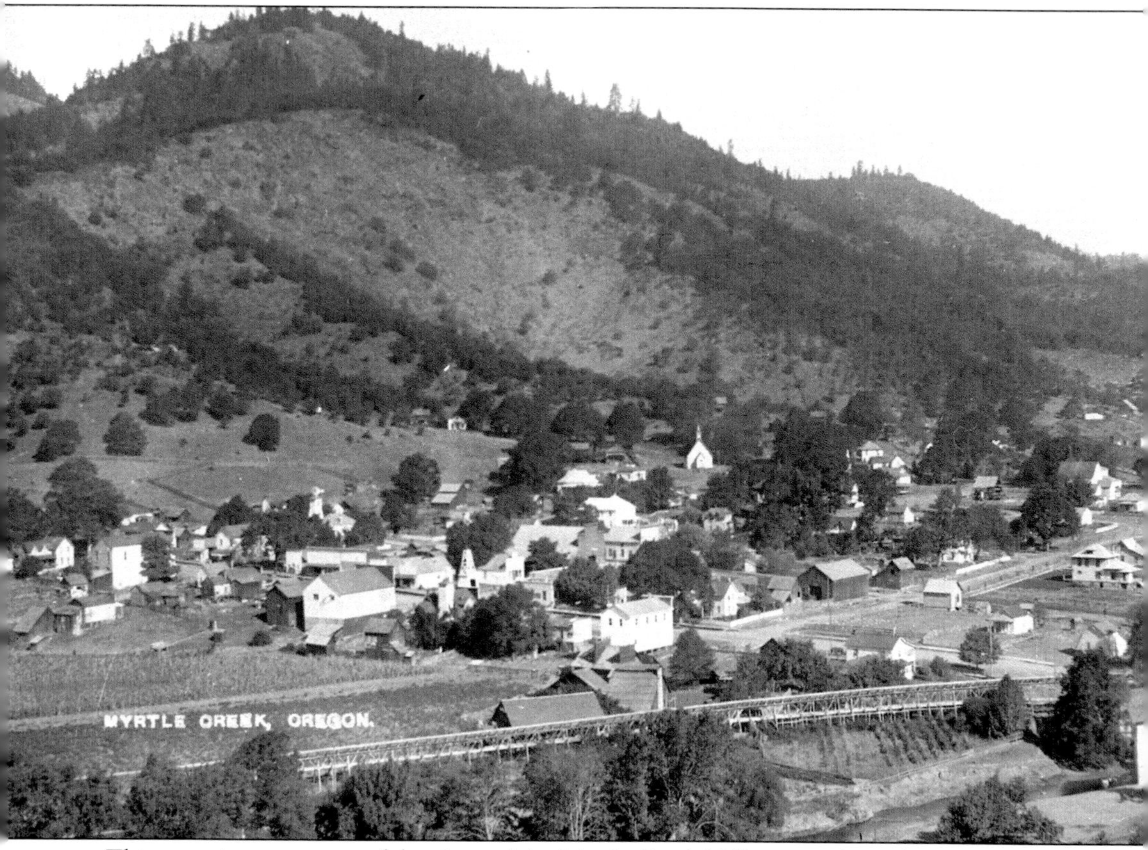

This sweeping panorama of the city and the Myrtle Creek Valley was taken after 1909, when the brick school on Division Street was built and before 1911, when the covered bridge across Myrtle Creek was replaced with a steel truss bridge. The school, at upper center on the right-hand page, is obscured by the grove of large oak trees in which it was built. The right foreground is filled

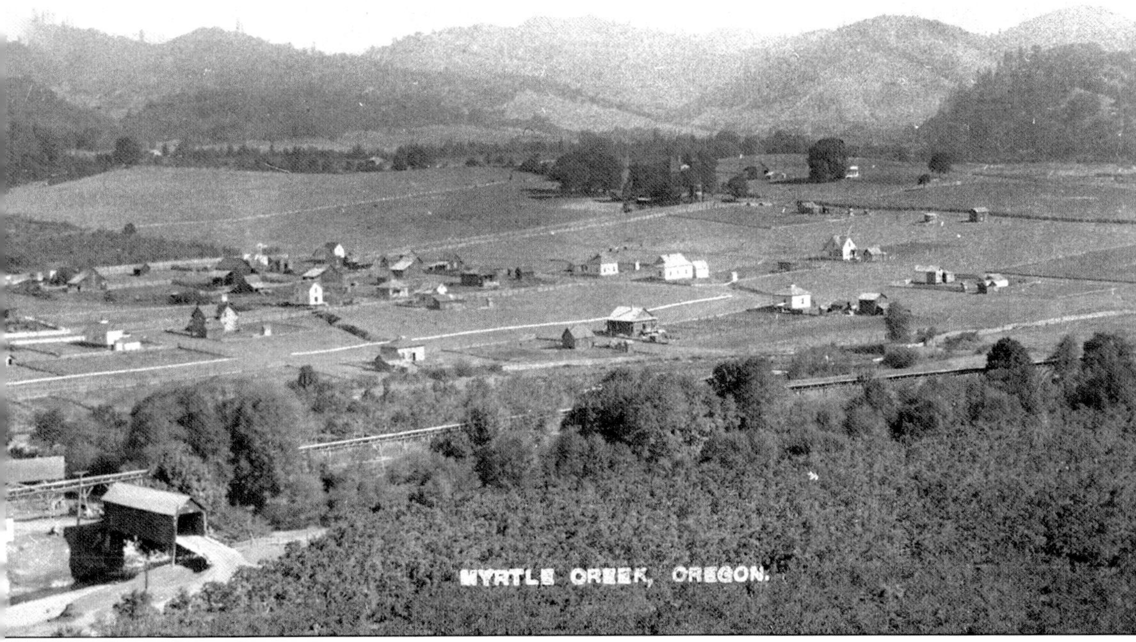

with John Hall's prune orchard, and at the left, where "Myrtle Creek, Oregon," is written on the photograph, John's son Todd has part of his field planted in corn. Over the top of the gristmill at bottom center, the Jackson House is under construction. Sixty years later, Hall's Western Auto would be built on the Jackson House site.

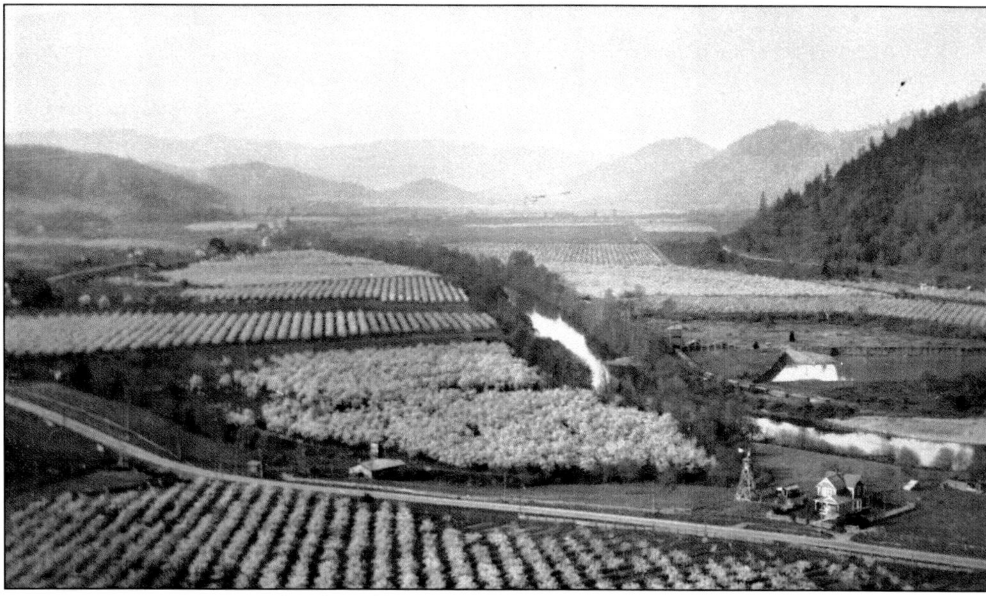

Missouri Bottom prune orchards in bloom surround the Adams house in the c. 1910 photograph above. By 1910, there were hundreds of acres of prune orchards and several prune dryers operating locally. The industry continued to grow, and in 1922 alone, eight more prune dryers were built. By 1925, there were 10,000 acres of prune orchards in Douglas County. In 1926, county prune orchards were assessed at $2.5 million, producing 22 million pounds of the dried fruit. Shown below is one of several local prune labels that were used, but this one is unique in that it is an artist's rendition of the photograph above.

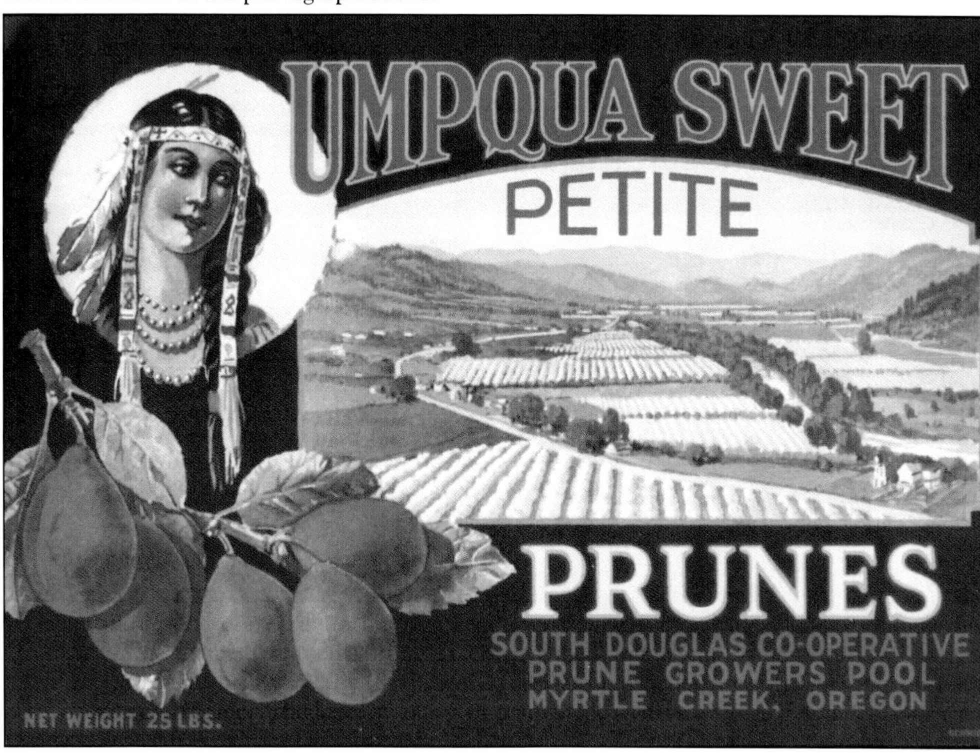

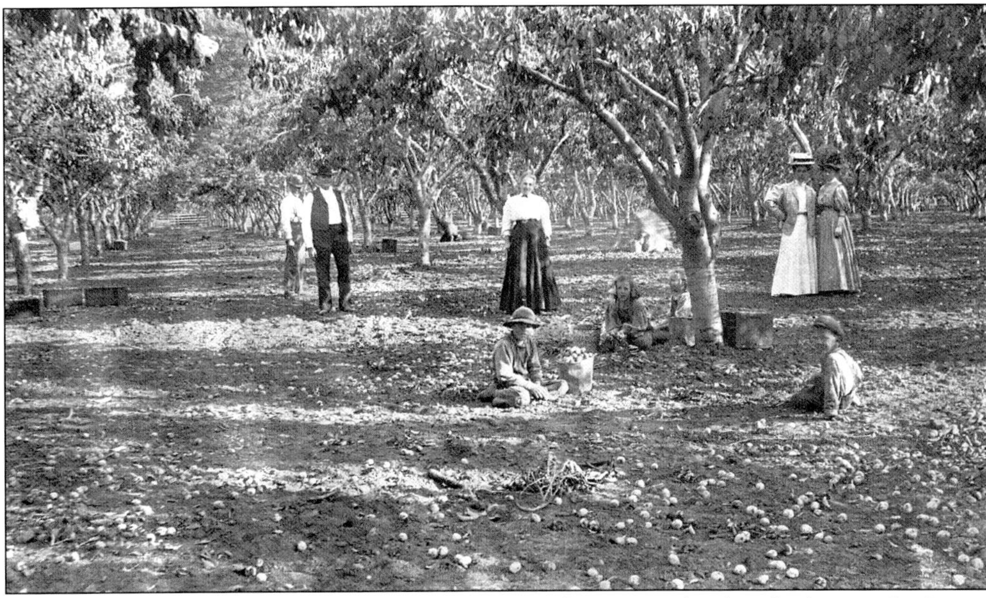

Family members and neighbors harvest the John Hall prune orchards, which were the first in the valley. For almost 40 years, prunes dominated the agricultural scene in the Umpqua Valley and provided work for many people. The opening of school was delayed in the fall until students were done with the harvest. For many, this was the only money available for school clothes and supplies.

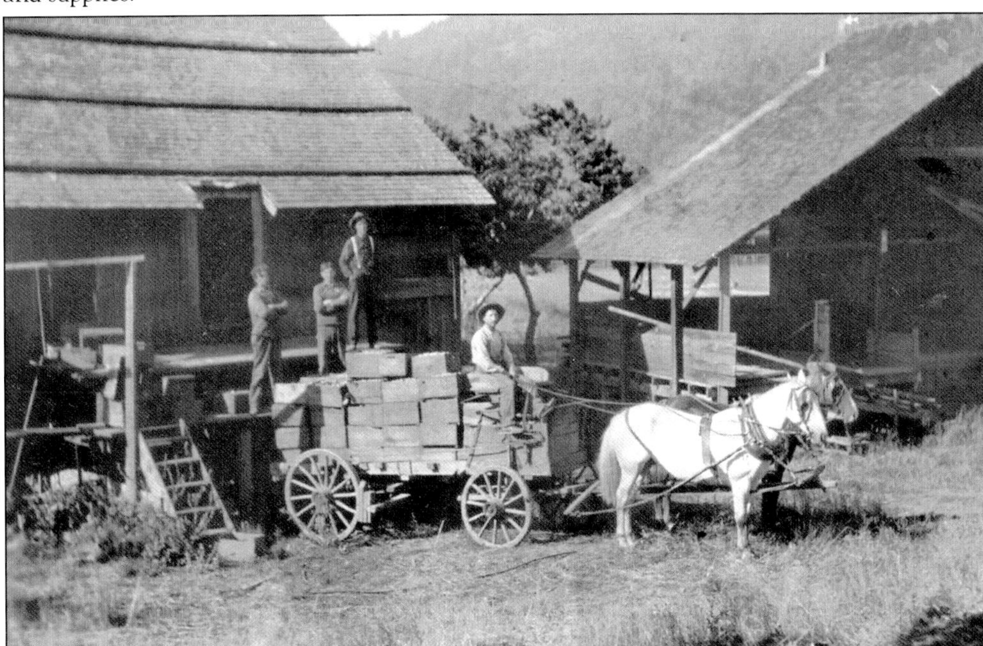

Loaded and ready for the trip to the train depot, Jesse Smith sits astride a load of crated prunes fresh from the dryer. In the fall, prune dryers provided ample employment for local workers. Not only were dryer crews needed to process, sort, and pack the fruit, others were contracted as woodcutters to supply the large amounts of cordwood needed to fire the dryers. (Courtesy of Lorelee Iles Dendauw.)

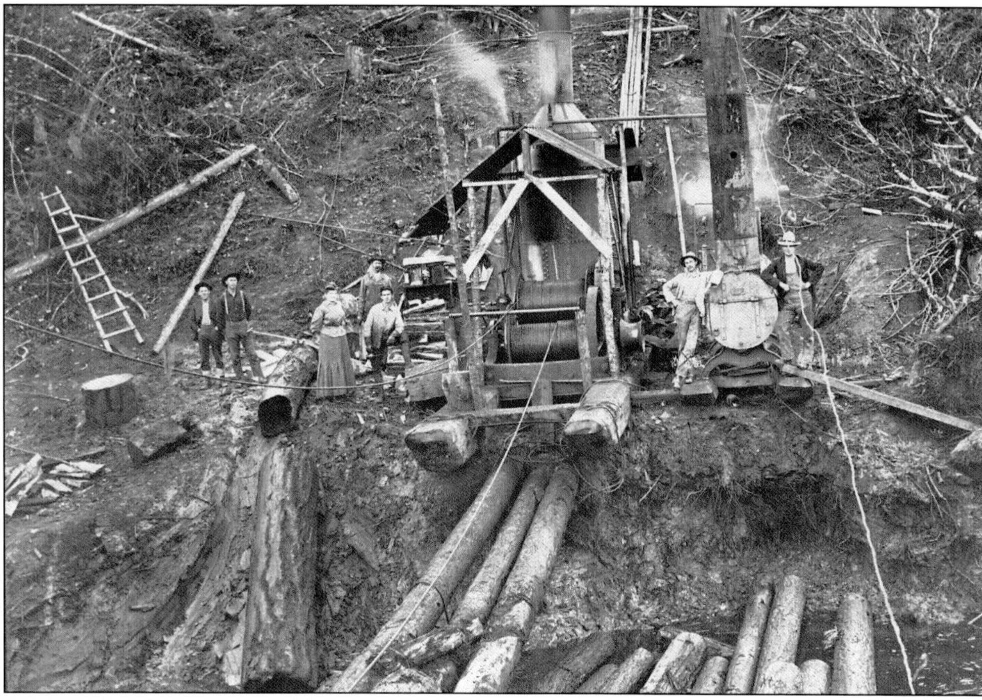

Once the local, easily accessible timber was harvested, other methods and machinery were needed to handle the trees. The invention of the steam donkey proved to be an important development that enabled the harvest of trees on steep ground. Capable of yarding large logs up steep hillsides, the steam donkey paved the way for the ensuing growth of the timber industry.

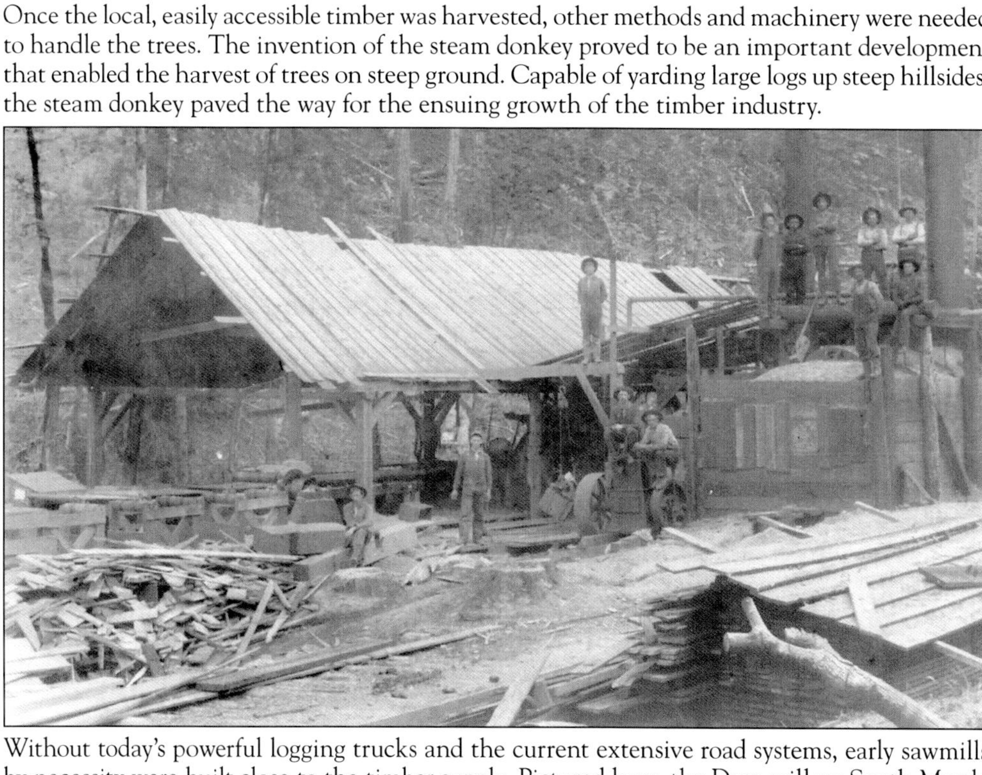

Without today's powerful logging trucks and the current extensive road systems, early sawmills by necessity were built close to the timber supply. Pictured here, the Dyer mill up South Myrtle is a typical early sawmill with crew members who probably lived within an easy ride or walk of the mill. (Courtesy of Dora Dyer.)

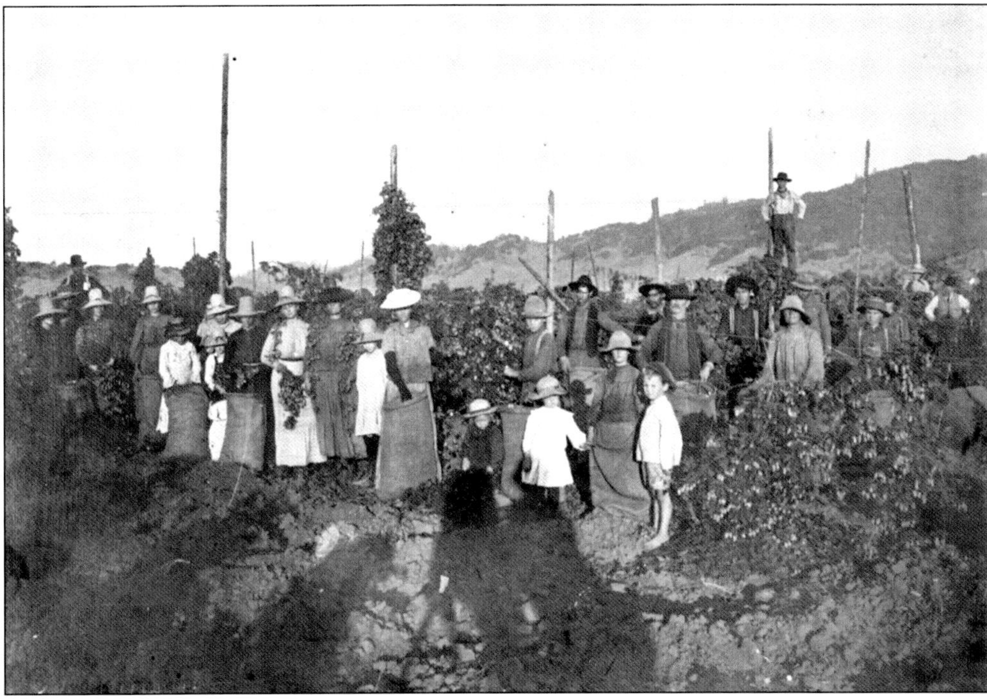

Hops were one of the first crops planted by the pioneers of the valley. Not that they were desperate for alcohol, but fermented beverages provided the safest form of liquid for everyone—children included—to drink. Here, Ed Weaver (left rear) poses his Missouri Bottom hop-pickers for the camera with the shadow of the photographer and camera in the foreground.

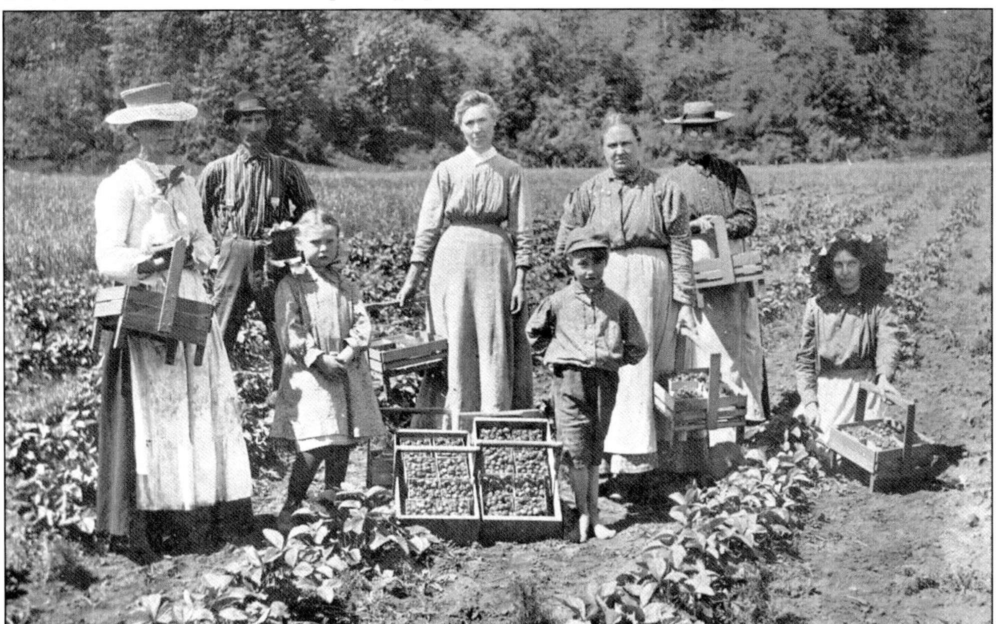

A group of pickers including Ed Strong (second from the left) and Mary Stelzier (right) pose with strawberries from a local patch. In *The Golden Leafs of Time*, many children recalled that spring strawberry picking provided money for school clothes and summer recreation.

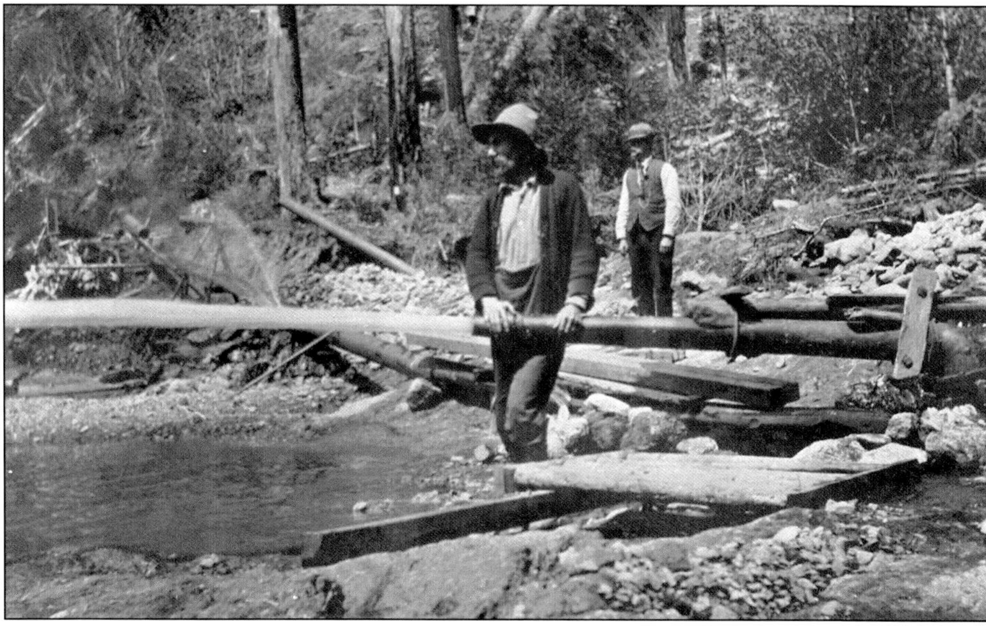

Hydraulic mining was the method of choice, and nearly all of the streams in the Myrtle Creek drainage were subjected to the erosive effects of the powerful nozzle called a monitor, or "Long Tom." Pictured in 1913, J.M. Martin works his Lee's Creek claim, hosing down the stream banks and streambed and washing the soil through a sluice box designed to capture gold.

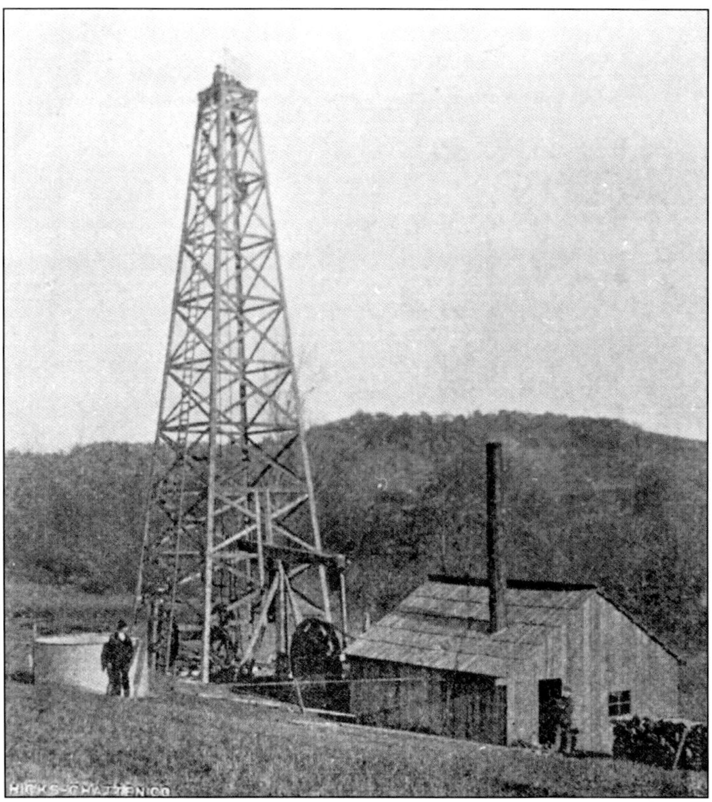

In December 1901, the Umpqua Valley Oil Company began drilling for oil near Myrtle Creek. At 1,700 feet, seepage was found, but a theft of tools and lack of funds forced the company to abandon the venture. The president was Malancthon McCoy, and A.C. Marsters was secretary. A banker, legislator, and mayor of Roseburg, Marsters promoted tourism in Douglas County. Marster's Bridge on the North Umpqua is named for him.

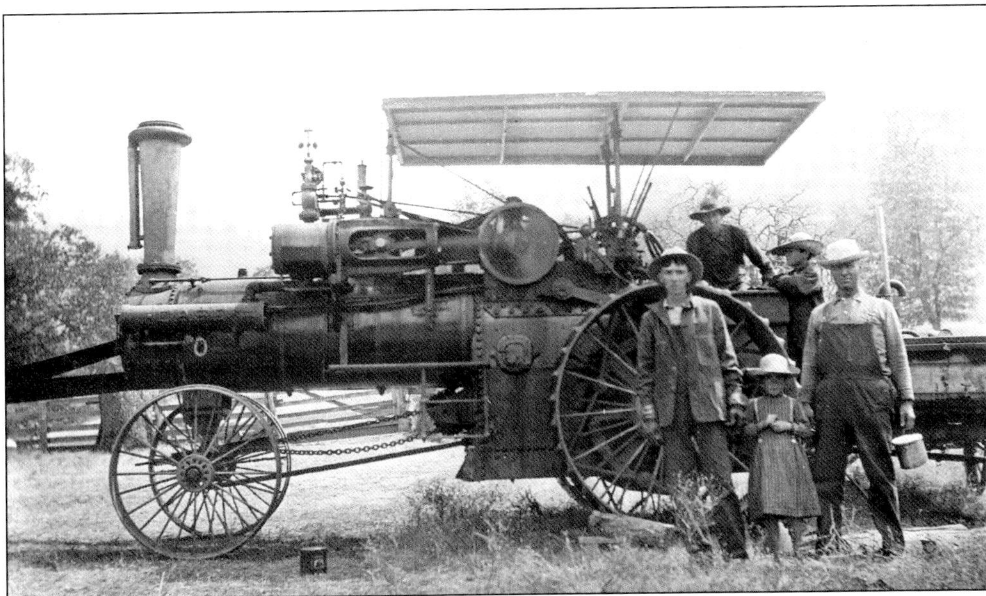

In 1904, the Pitts Ranch, located six miles north of town, had plenty of use for this ponderous piece of equipment. The invention of the steam tractor changed the face of agriculture for those who could afford it. Not only did it make plowing the fields faster and easier, the tractors also provided power to run irrigation pumps, threshing machines, portable sawmills, and various other items of farm machinery. (Courtesy of Al Helgeson.)

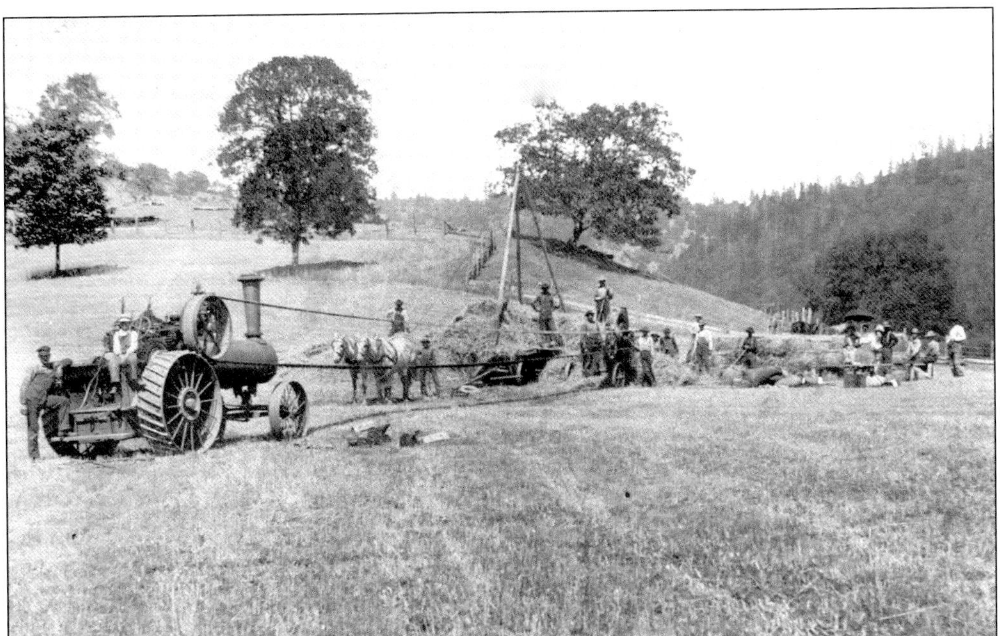

This c. 1915 photograph shows Barry Jones and a haying crew working a local ranch. Barry had a steam tractor, a threshing machine, and a hay baler. During the harvest season, he provided farmers the equipment needed for harvest. Note how the long belt kept the tractor away from the work. If it got too close, the boiler fire could ignite the chaff dust, causing an explosion.

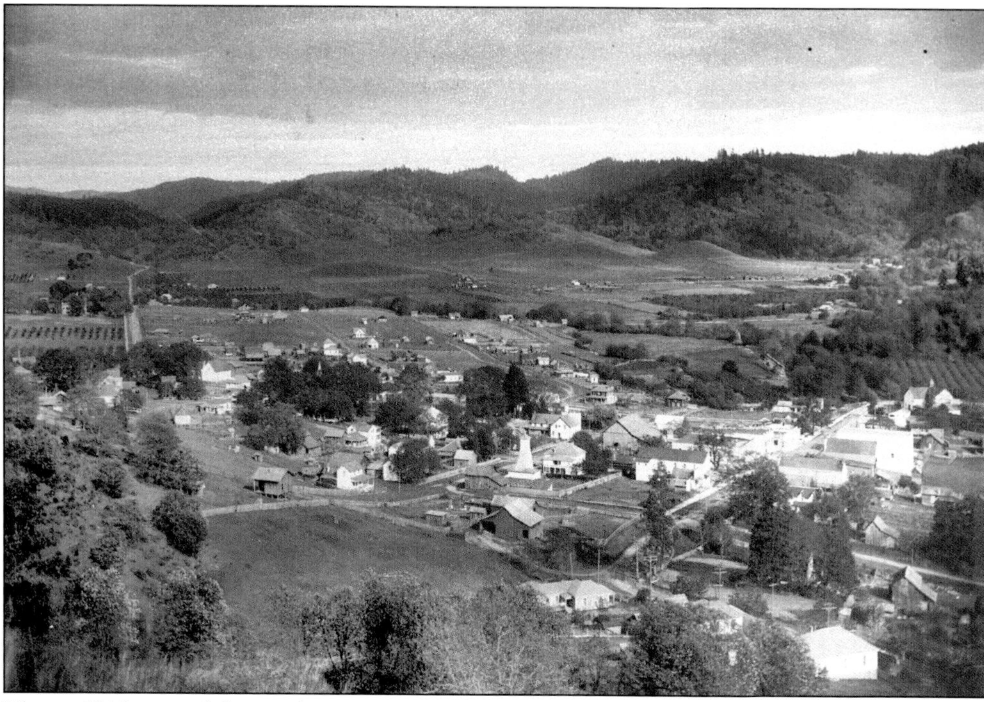

This c. 1915 view of the city has some interesting features. Dole Road is still the north entrance to town, but Fourth Avenue is well used, indicating that the Weaver Road Bridge is in place. Just a remnant of the Johnson mill flume remains, but the 60-year-old millrace is still evident. At Third Avenue and Oak Street, Charles White's house stands with a windmill behind it.

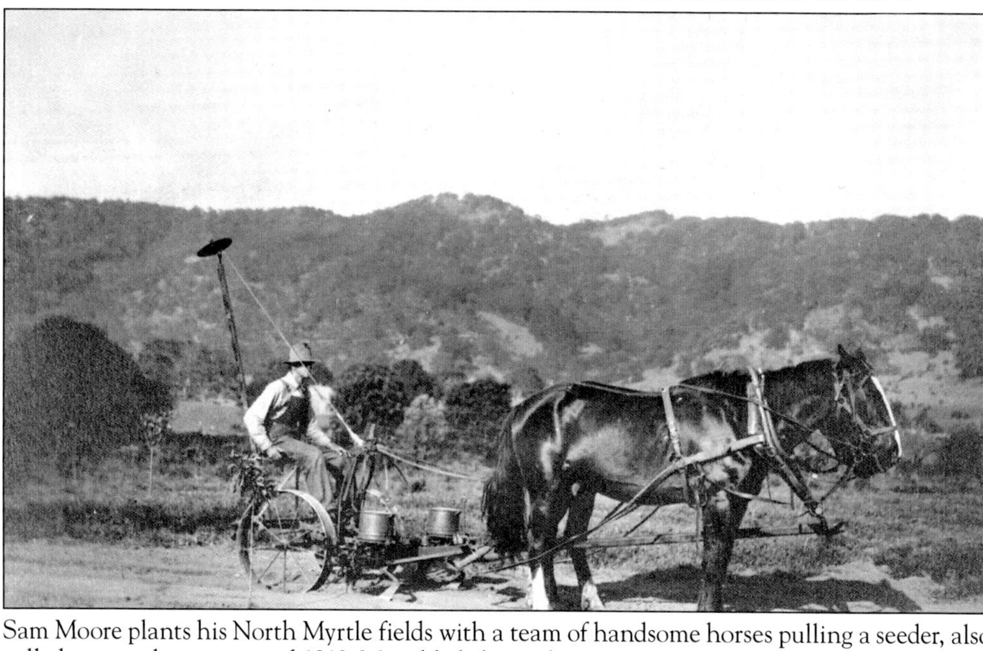

Sam Moore plants his North Myrtle fields with a team of handsome horses pulling a seeder, also called a corn planter, around 1910. Most likely he is planting corn, or maybe beans. The tall arm over his head was dropped to mark even planting rows. In the foreground is a freshly planted sapling, probably part of a tract of prunes he had just set out.

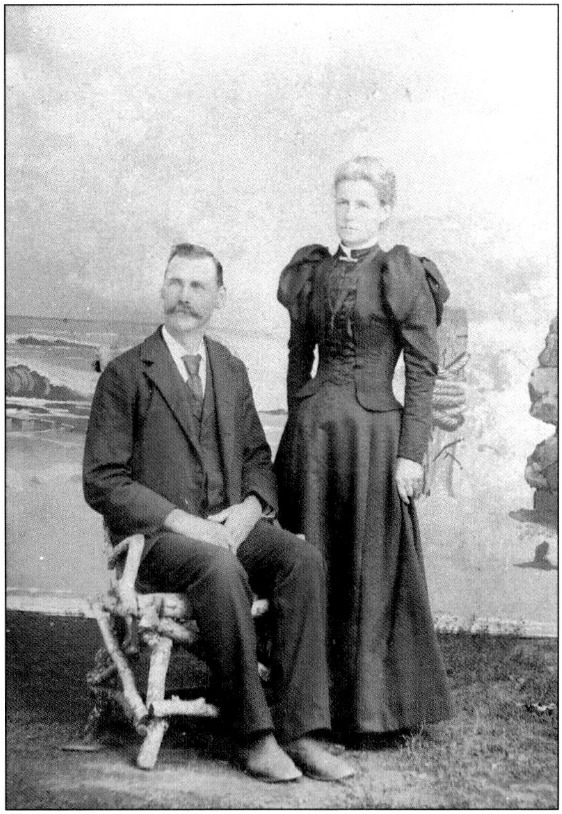

On the right is the wedding photograph for the July 7, 1867, marriage of James S. Dunnavin (1845–1926) and Susan Sara Browning (1848–1919). Susan's uncle was the renowned John Browning, owner and founder of the Browning Arms Company. James was an early freighter who in the 1870s built a home called Mulberry Glen on their Lee's Creek ranch. Retiring to Division Street in town, he built a distillery and shipped brandy to various markets. Below, his son James E. Dunnavin proudly poses with one of his mules. He bred and sold mules from property between Division Street and Riverside Drive that became known as "Jackass Flats."

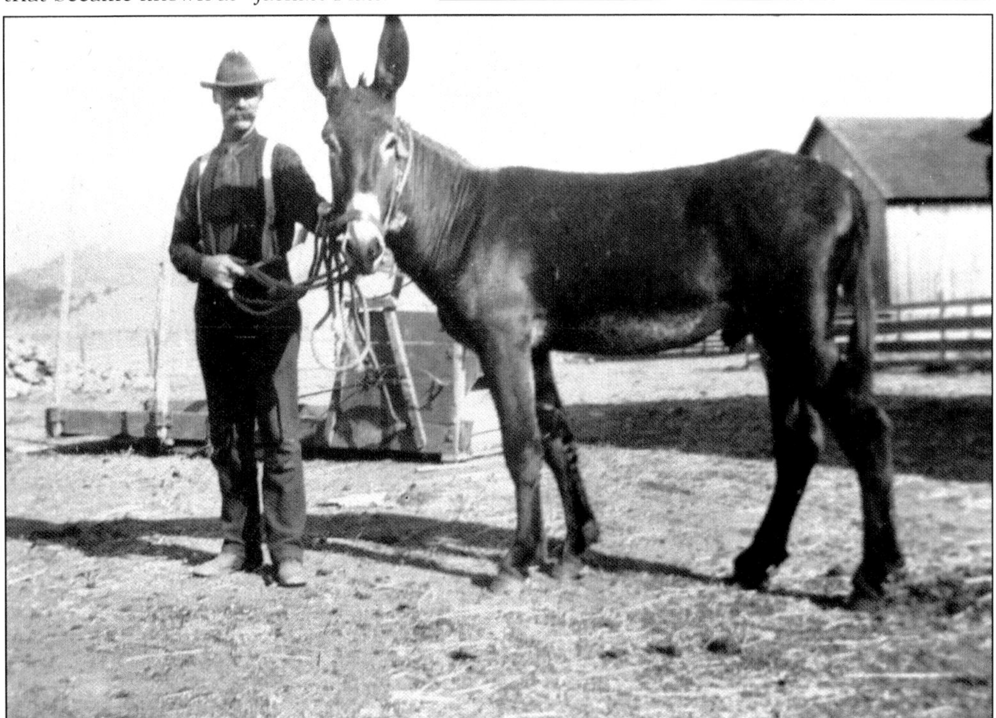

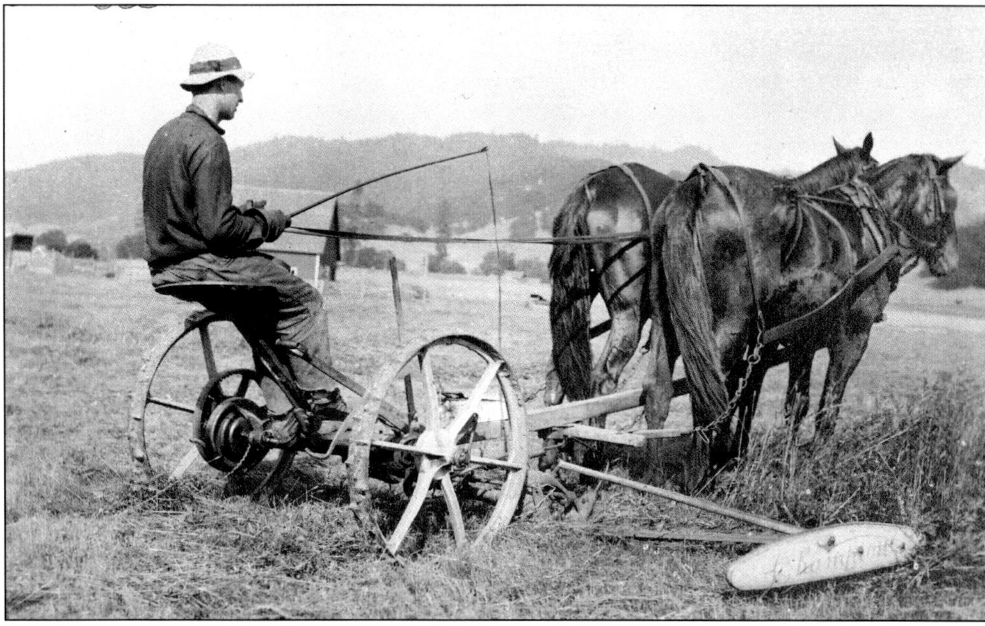

In the c. 1914 photograph above, Ernest Bowman keeps a whip at the ready to urge on the two-horse team pulling his Champion sickle-bar mower. Another c. 1914 photograph below shows J.F. Barker's Myrtle Creek business, where it is likely that Ernest purchased the mower. Advertising for Studebaker, Mitchell, and Champion, Barker kept two buildings to house his inventory. Barker is listed in the 1910 Myrtle Creek census, and in a 1903 Roseburg newspaper are advertisements for his stores in Roseburg and Myrtle Creek. By some quirk of fate, all knowledge of this business and its location has been lost in the haze of history.

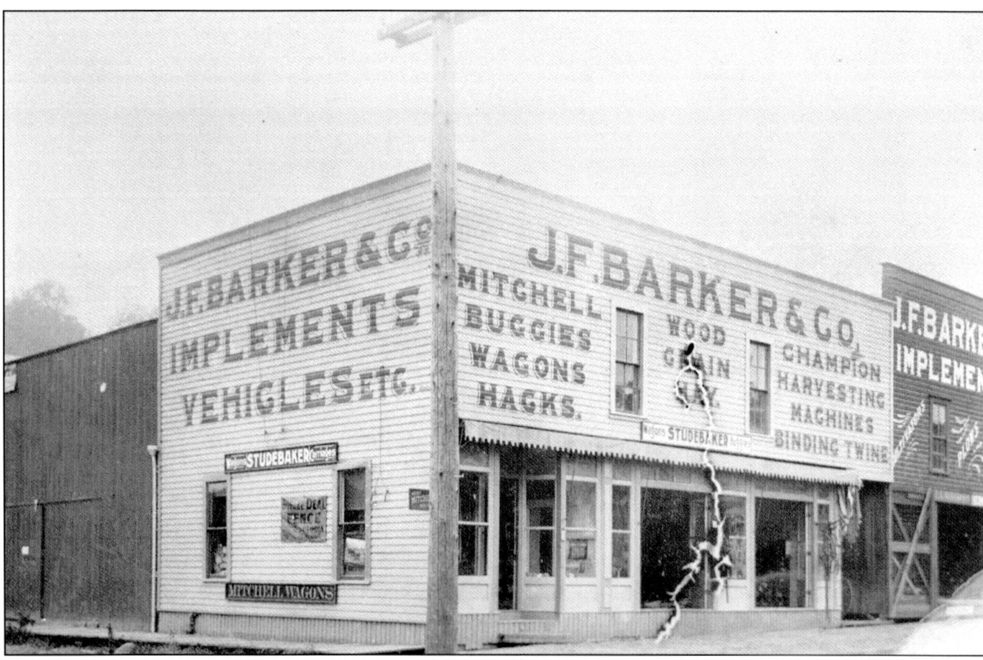

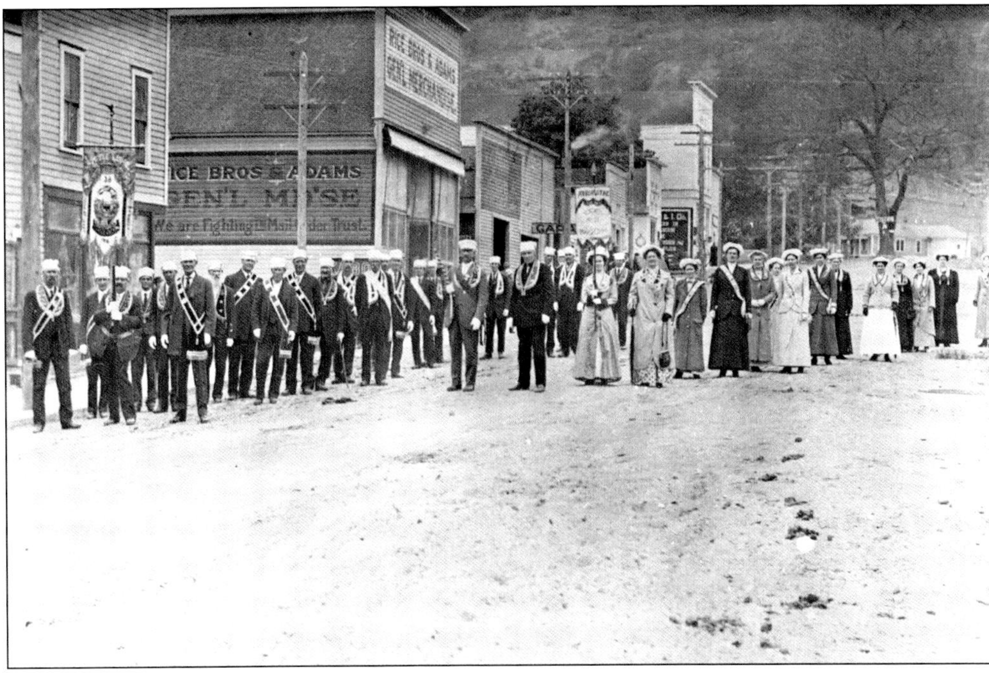

In this 1912 photograph, the members of the annual IOOF parade pose in the middle of Myrtle Creek's manure-strewn Main Street. Citizens participating in the festivities include merchant Nathan Selig, distiller James Dunnavin, and cobbler Dick Harlow.

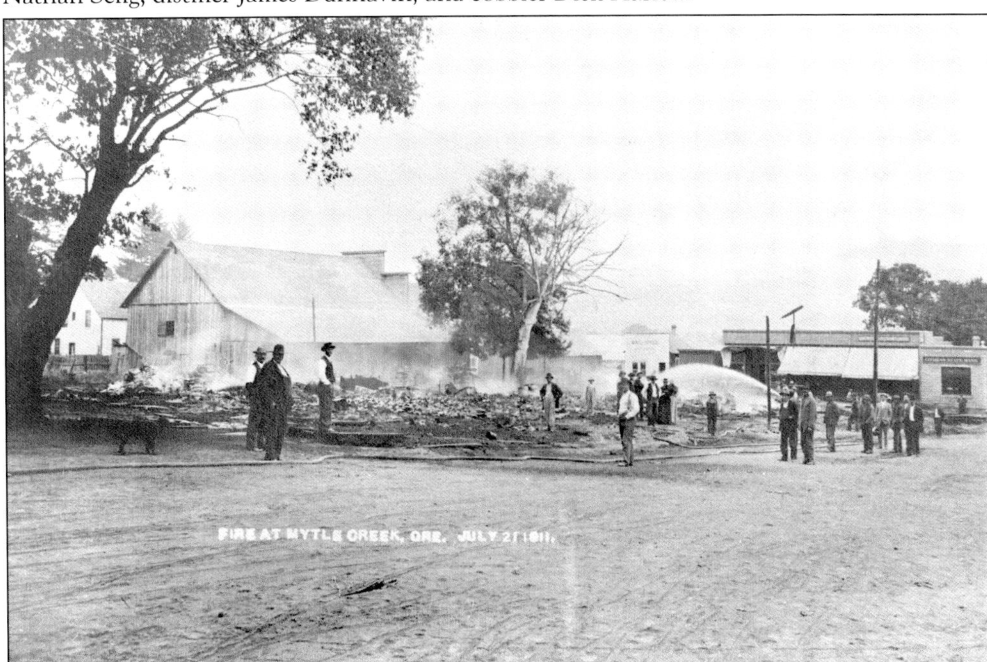

This fire on Main Street between Second and Third Avenues burned one building and a windmill on July 21, 1911. The large building on the left is Weaver's Wagon Shop, and behind it is the Central Hotel. The Moore building and the Weaver building were constructed on this site in 1912. Both buildings still exist; the top floor of the Weaver building became Myrtle Creek's first hospital.

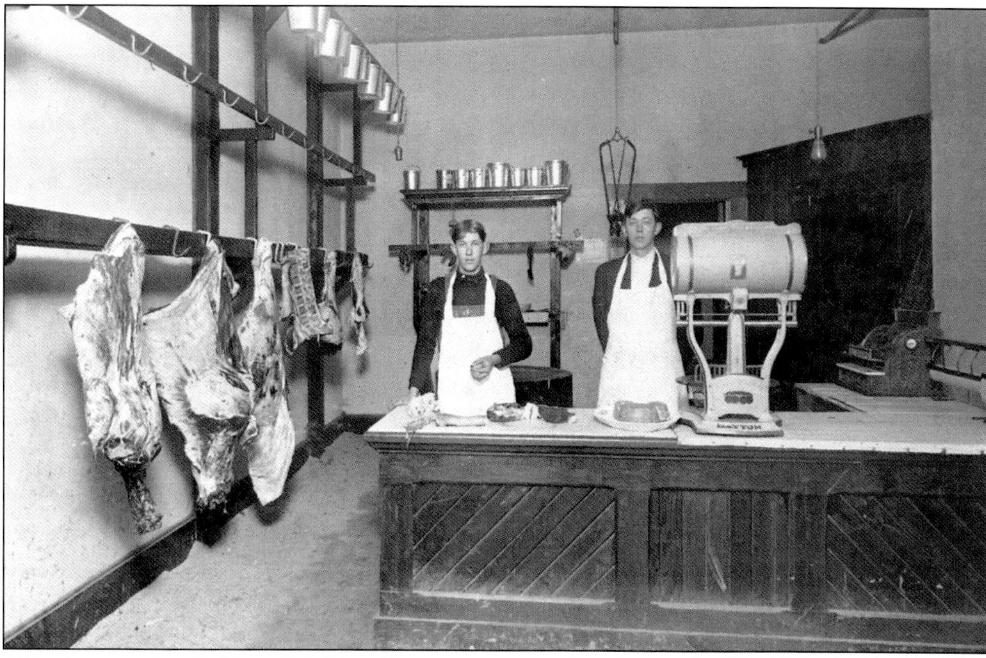

Older brother Art (left) and William Dyer are portrayed in this 1913 photograph of an unknown Main Street butcher shop. It featured Dayton scales and an ornate National cash register, with lard pails and sides of beef hanging from hooks. With a butcher-paper dispenser and cuts of meat on the counter, this shop appears neatly kept and well equipped.

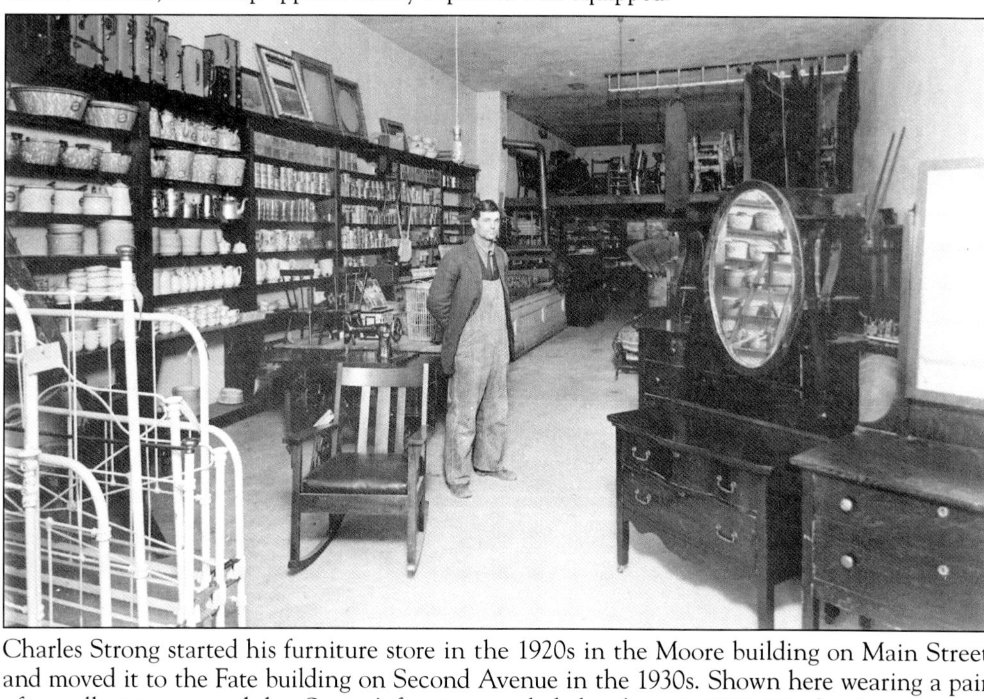

Charles Strong started his furniture store in the 1920s in the Moore building on Main Street and moved it to the Fate building on Second Avenue in the 1930s. Shown here wearing a pair of overalls, it is reported that Strong's furniture included caskets. In 1943, His daughter Zenobia Strong McGinnis and Dr. Verne Adams started the first Myrtle Creek Hospital on the top floor of the Weaver building.

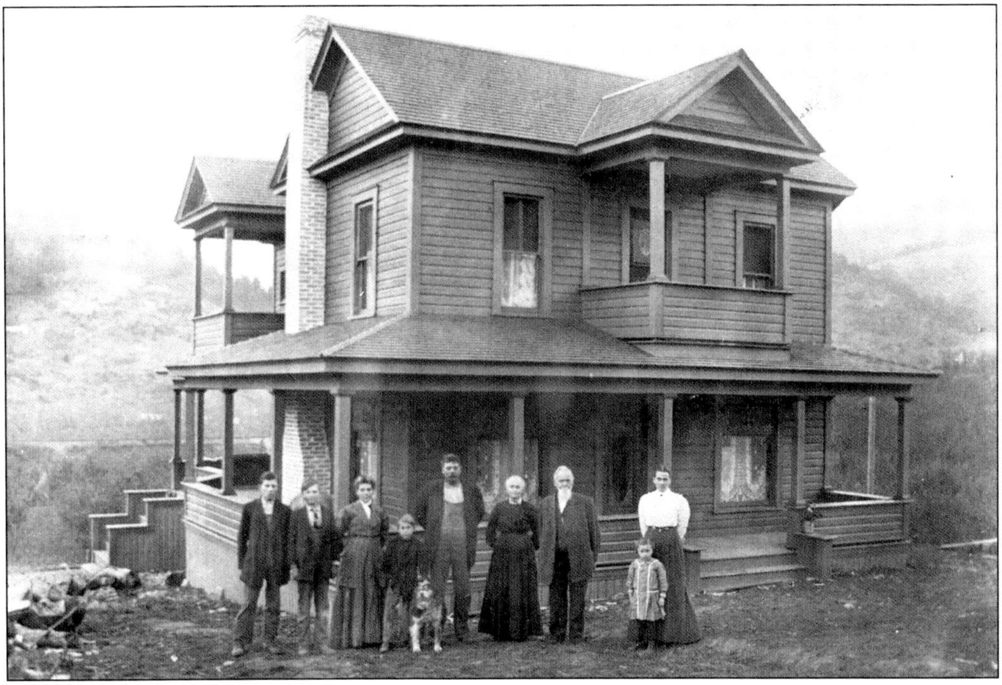

Starting with a blind horse and 75¢, Joel Harris parlayed this meager beginning into a prosperous life and this beautiful home. Built on the west corner of Western Avenue and Pacific Highway, the house overlooked his prune orchard and the Western Hotel at the train depot. His daughter LaVonna married Marion Kusler; for over 25 years, they operated Kusler's Super Mart on Main Street.

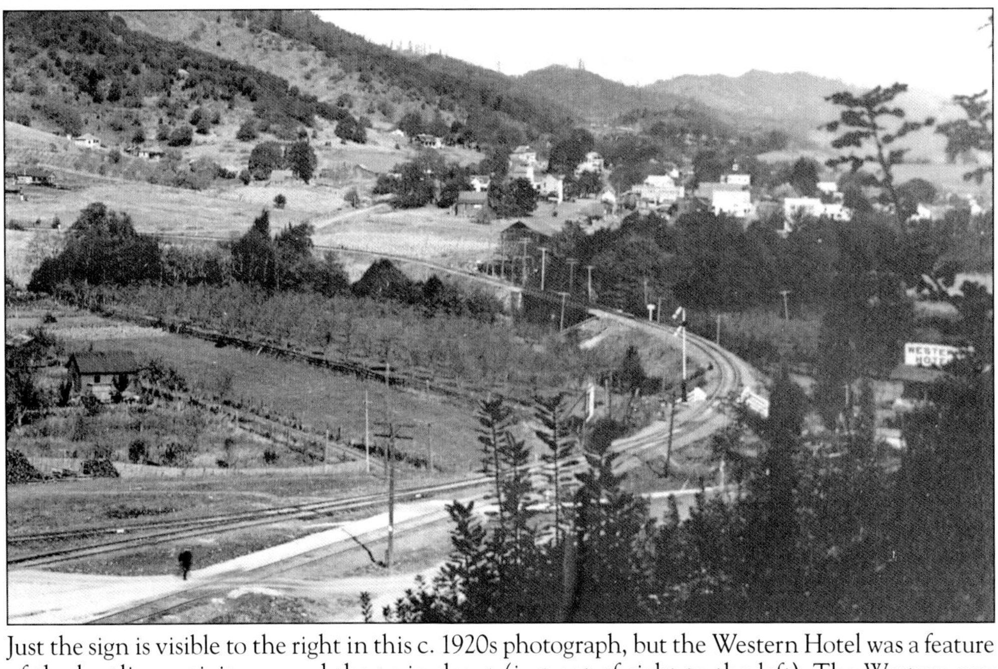

Just the sign is visible to the right in this c. 1920s photograph, but the Western Hotel was a feature of the bustling activity around the train depot (just out of sight to the left). The Western was operated by the Harris family and shared the depot area with a bulk oil plant, packinghouses, a stockyard, and sawmills.

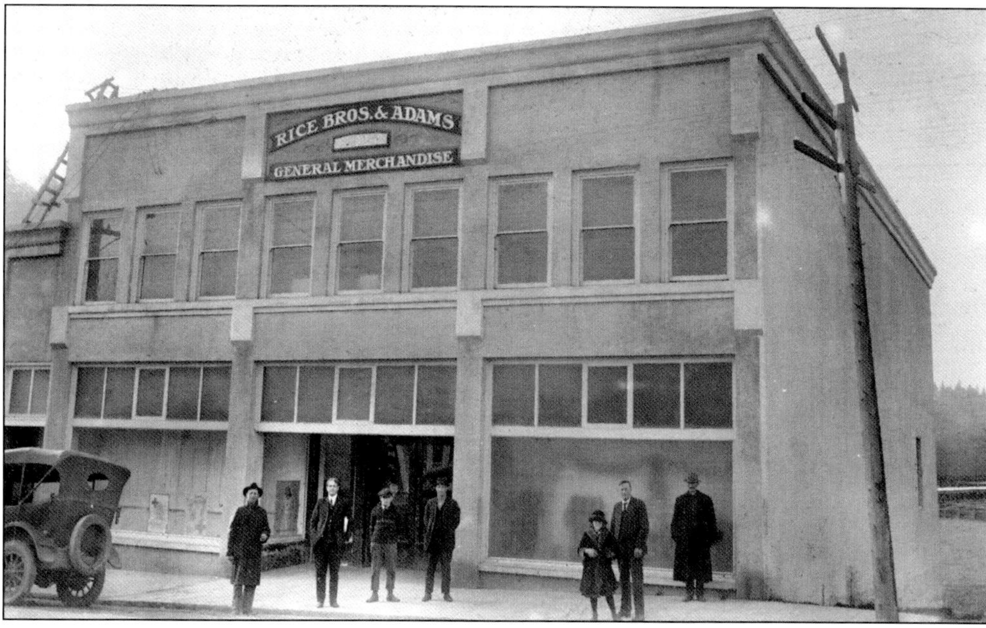

In 1915, Rice Bros. and Adams built this monolithic, poured-cement building, utilizing labor from citizens working off their grocery bill. The new store became a popular public gathering place, hosting daily gossip and heated political debates. An open barrel of ginger snaps ensured that lingering children were seen and not heard. Over the years, the building, which still exists, has housed a wide variety of businesses.

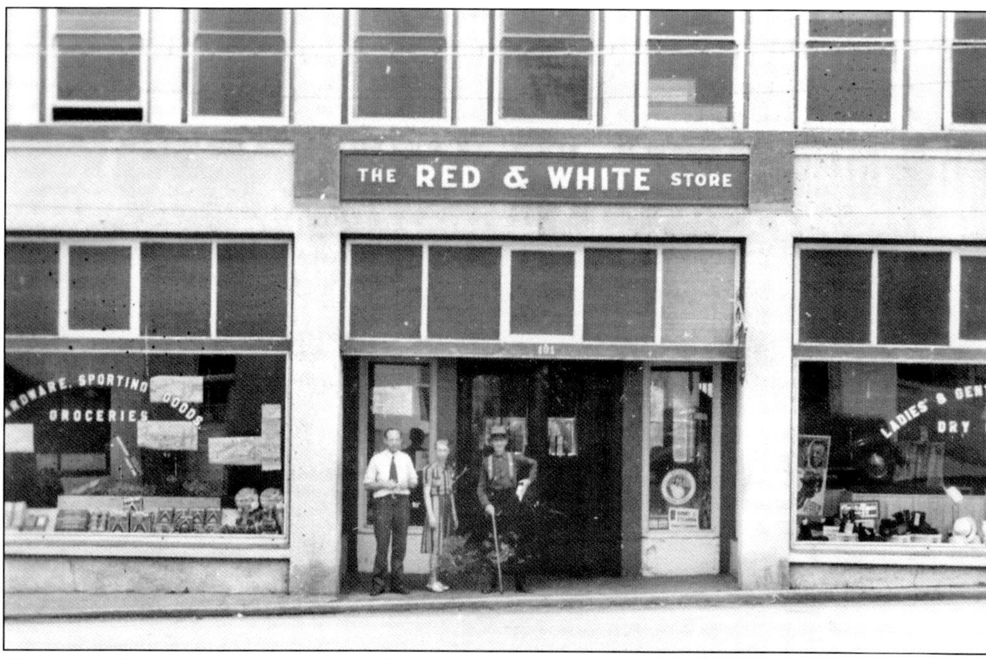

This c. 1920s photograph shows a different incarnation of the Rice building. The Red & White Store was a grocery store brand that Rice Bros. and Adams once carried. Along with groceries, they carried a variety of feed and farm supplies, hardware, sporting goods, stationery, toiletries, and clothing. The Rice Bros. and Adams store was Myrtle Creek's "one-stop shopping center."

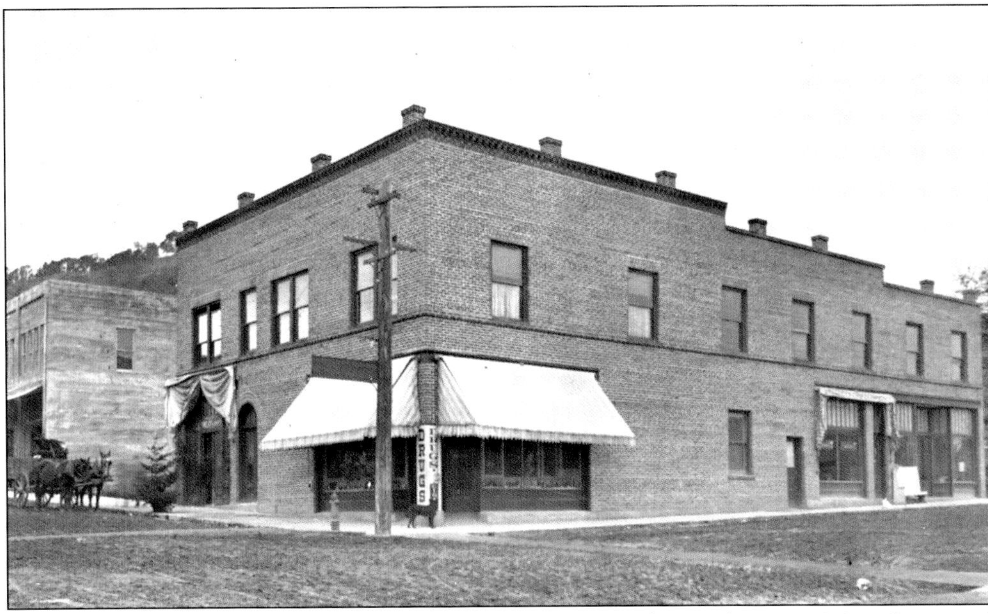

The Weaver brick building was constructed around 1912 with storefronts on Main Street and Second Avenue with Myrtle Drug occupying the corner. The second story was a hotel; in 1943, it was converted to use as a hospital. J.E. Creason was the first owner of Myrtle Drug, and Ted Rice purchased it in 1936. In the 1930s, Shirley Temple and her mother stopped at the fountain for a soda.

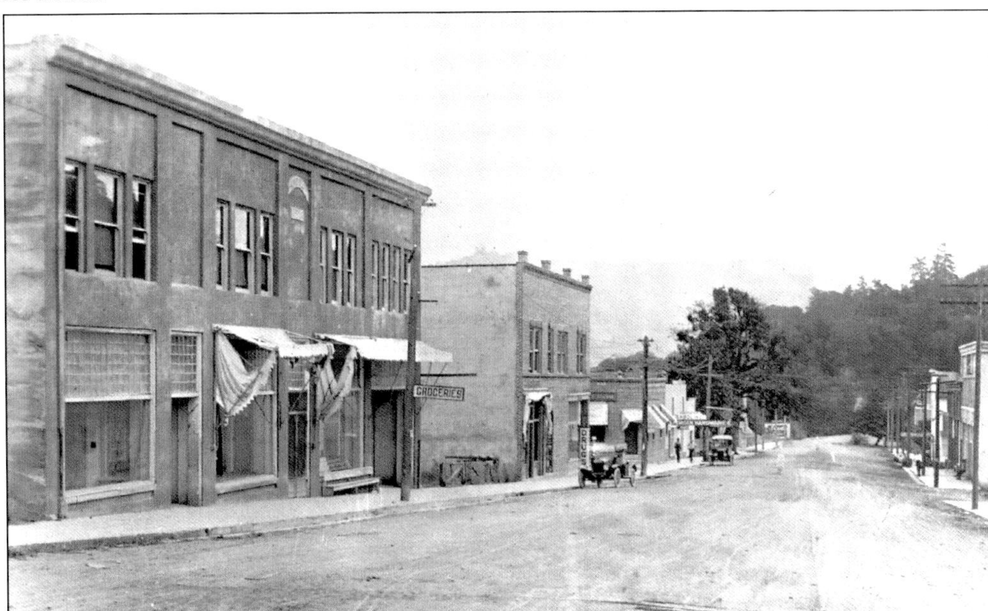

This 1920s-era Main Street photograph shows the Moore (left) and Weaver buildings. Constructed around 1912 of cast-in-place concrete, the first floor of the Moore building was originally divided into three storefronts. At one time, a theater and the telephone exchange were located on the upper level, which is now the IOOF hall. Kusler's Market occupied the main level from the early 1930s to the late 1950s. (Courtesy City of Myrtle Creek.)

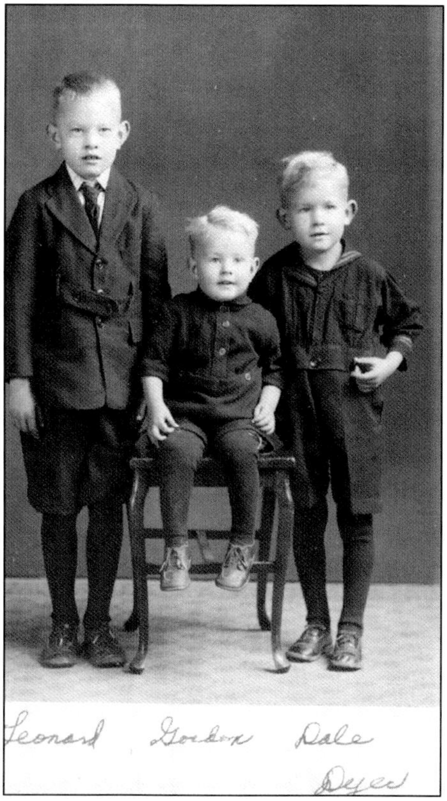

From left to right, Leonard, Gordon, and Dale Dyer were the sons of Goldie and Almira Clack Dyer and the grandsons of Moses True Dyer, one of Myrtle Creek's first settlers. All three graduated from Myrtle Creek High and fought in World War II. Dale was killed in a 1947 logging accident, Gordon was a logging loading engineer and high-climber until a disabling injury in 1960, and Leonard owned a local insurance business. (Courtesy of Dora Dyer.)

In 1897, Mary Gallop purchased land on Orchard Drive from Ferd Gabbert for $200 and, with husband Hiram, erected this Queen Anne home around 1910. It was the first house in town built from a Sears, Roebuck and Co. ready-to-assemble house kit. These kits were sold from 1908 to 1940 and were shipped by rail. Purchased by Goldie and Allie Dyer in 1933, this house is still in use. (Courtesy of Dora Dyer.)

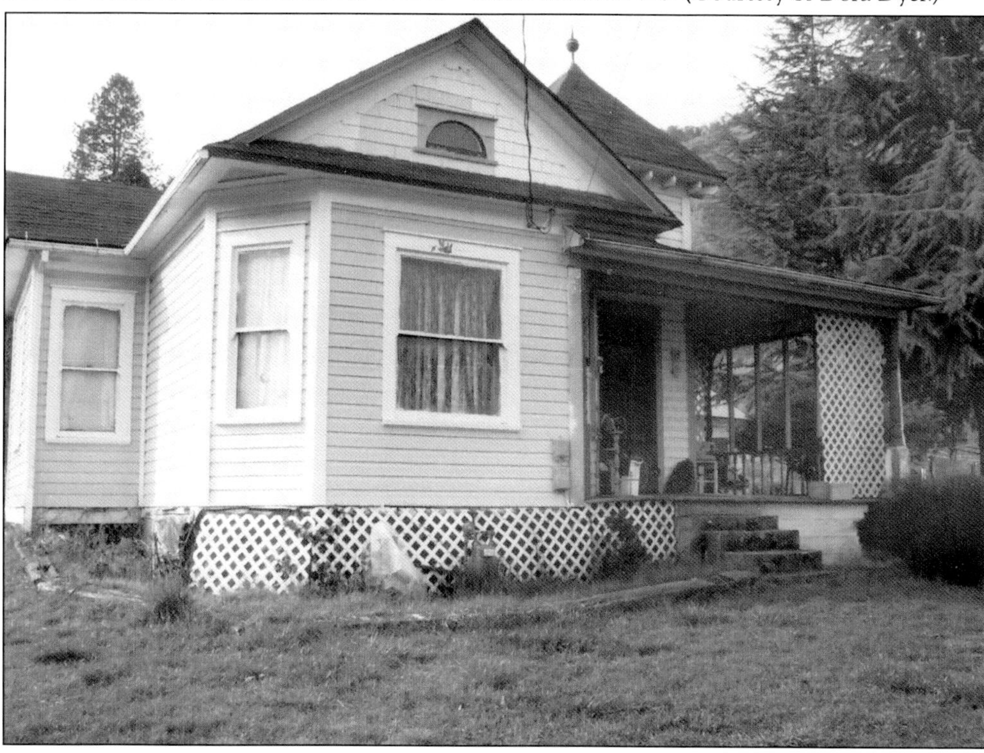

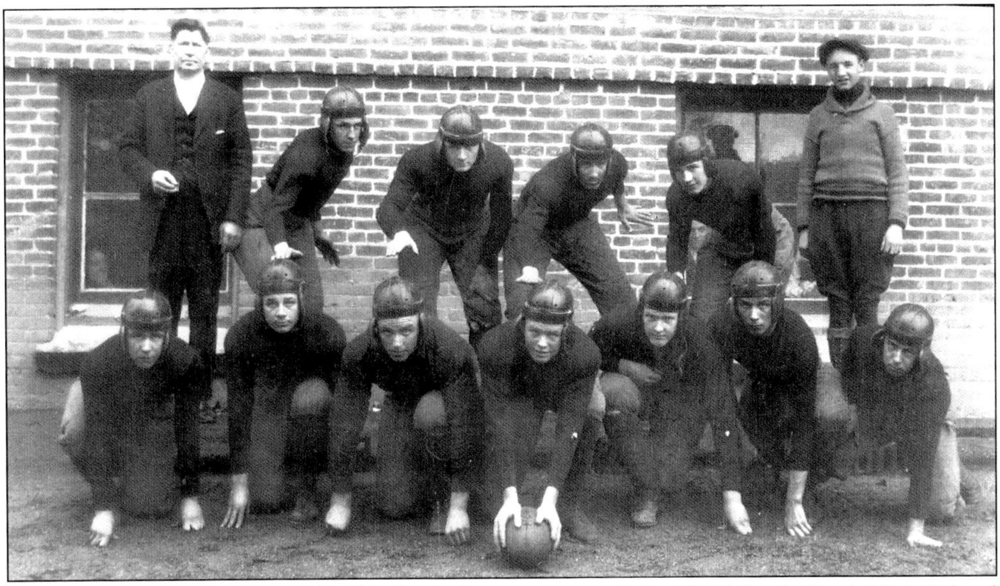

These rugged-looking men clad in leather helmets comprised the 1923 Myrtle Creek High School football team. From left to right are (first row) Lavern Merritt, Ted Shirtcliff, Alva Bowman, Ted Rice, Milton Dement, Estle Paris, and Paul Froehlich; (second row) Principal J.T. Rice, Floyd Baker, Joe Rice, Earl Way, Lloyd Ausmus, and Olaf Newton.

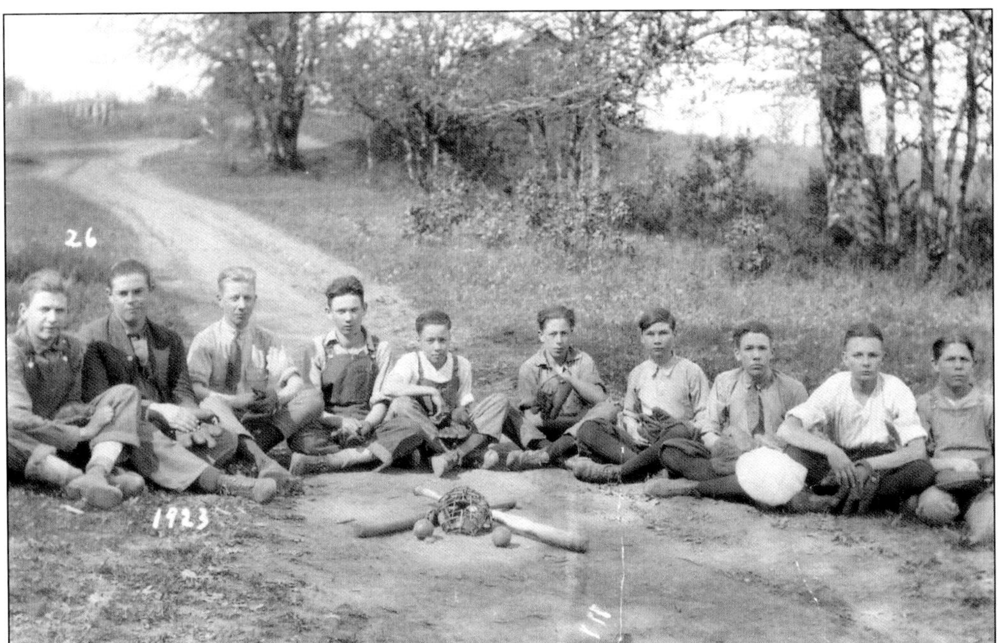

Spread across Division Street past the school and posing with a display of bats, balls, and mitts is the 1923 Myrtle Creek High School baseball team. The only identified player is Howard Brownson, fourth from the left. (Courtesy of Martha Brownson.)

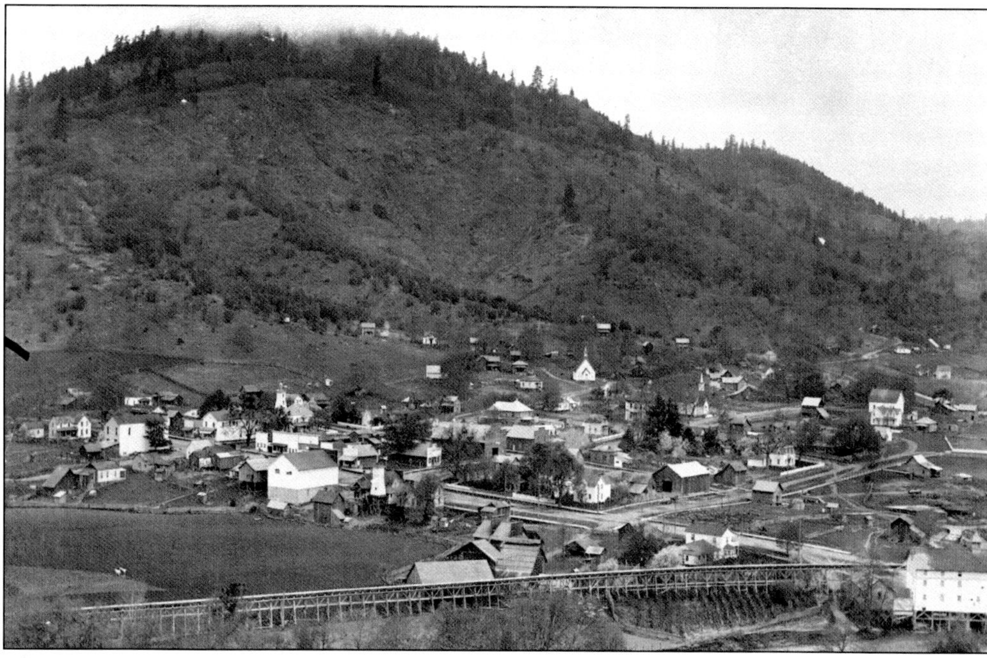

This 1907 panorama of Myrtle Creek, taken by Medford photographer F.H. Hull, shows the Johnson Mill lumber flume running across the bottom to the sorting facility at its railroad siding. Also barely visible on the lower left are two white spots that, under enlargement, proved to be two white horses pulling a plow, probably operated by Todd Hall.

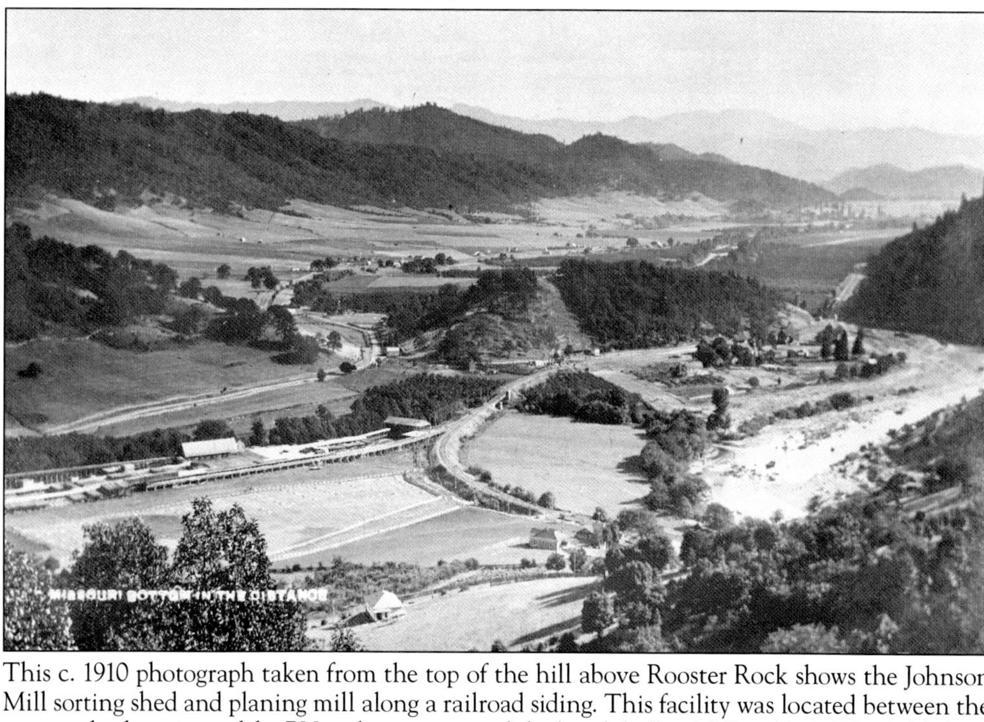

This c. 1910 photograph taken from the top of the hill above Rooster Rock shows the Johnson Mill sorting shed and planing mill along a railroad siding. This facility was located between the present-day locations of the RV park restroom and the band shell in Millsite Park. The field along the railroad siding shows that Todd Hall has shocked up about a third of his hay crop.

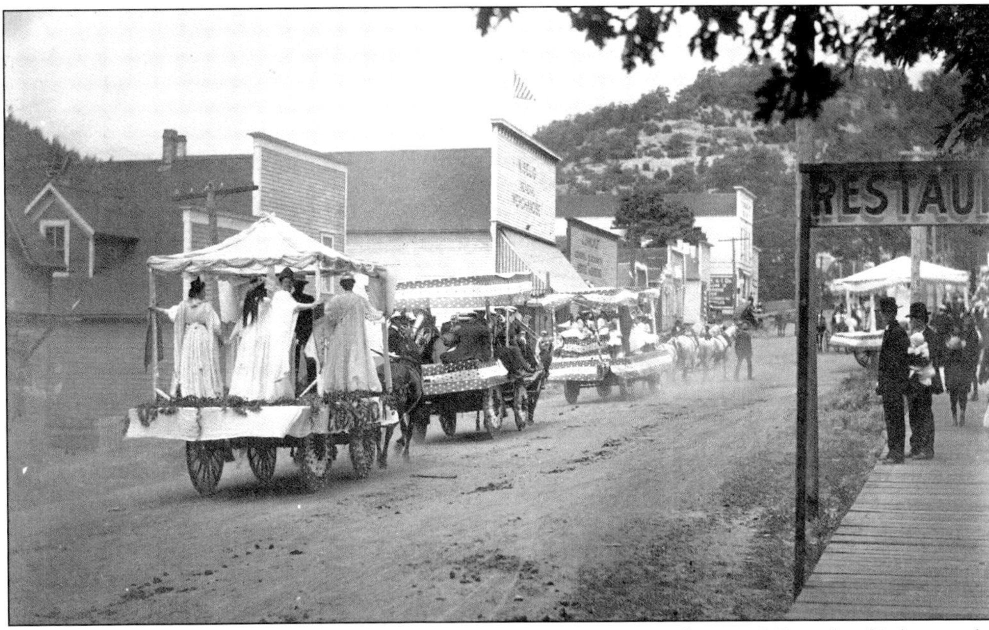

Decorated horse-drawn floats make their way up Main Street in this 1915 Fourth of July parade. The large storefronts on the left include those of Selig, Anlauf, and Hunsaker. Straddling the boardwalk between First and Second Avenues is a sizable "restaurant" sign that must have been a popular spot with travelers. (Courtesy of City of Myrtle Creek.)

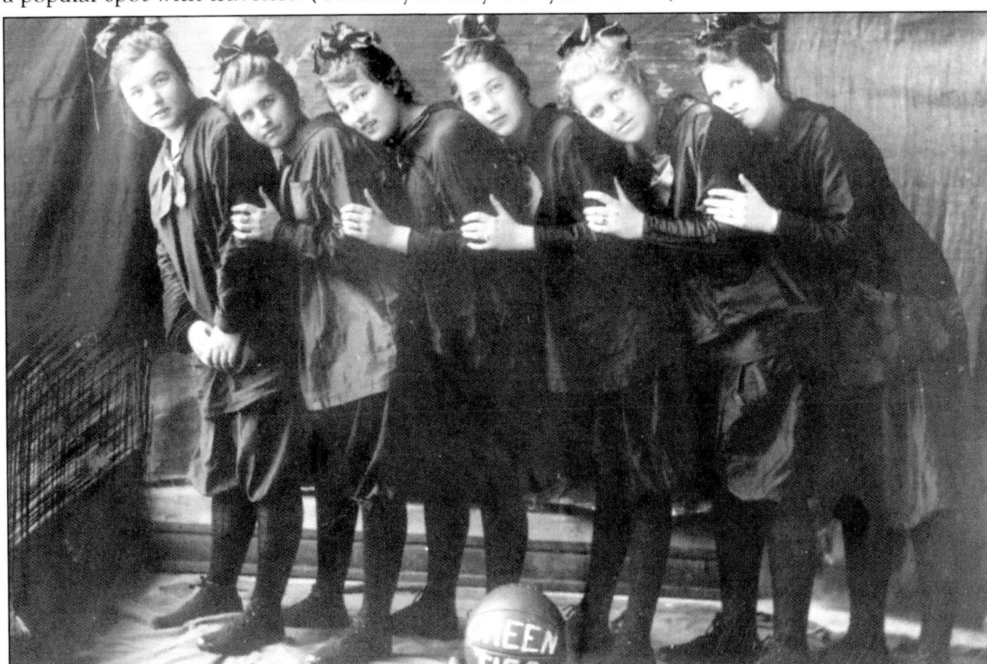

In 1916, the Green Ties, the girls' basketball team, represented Myrtle Creek High School. These winsome young ladies, listed with their future married names in parentheses, are, from left to right, Virgie Lathrup (Bushnell), Allie Clack (Dyer), Verena Way (Puckett), Veva Dyer (Wimer), Ruby Russell (Thomas), and Vivian Buell (Laurence). (Courtesy of City of Myrtle Creek.)

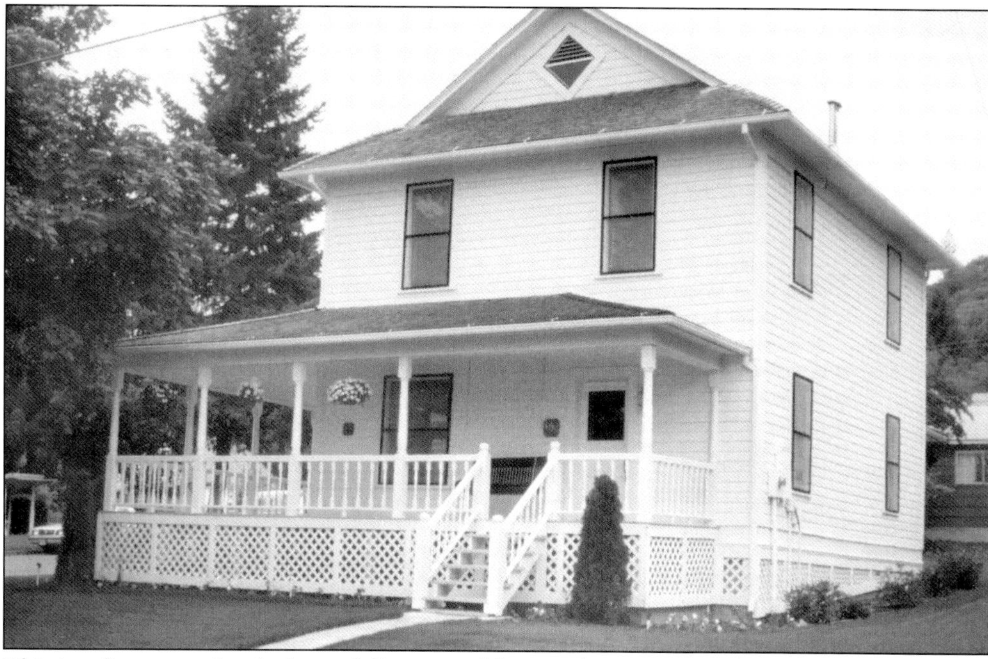

This is a fine example of a beautifully restored historic home. Located on Second Avenue and originally constructed around the turn of the 20th century, it once belonged to a man named McGee who was a blacksmith. His son Charles McGee plied the same trade and had a shop in Myrtle Creek. It is not known if Charles also lived in this house as an adult. (Courtesy of Florence Chaney.)

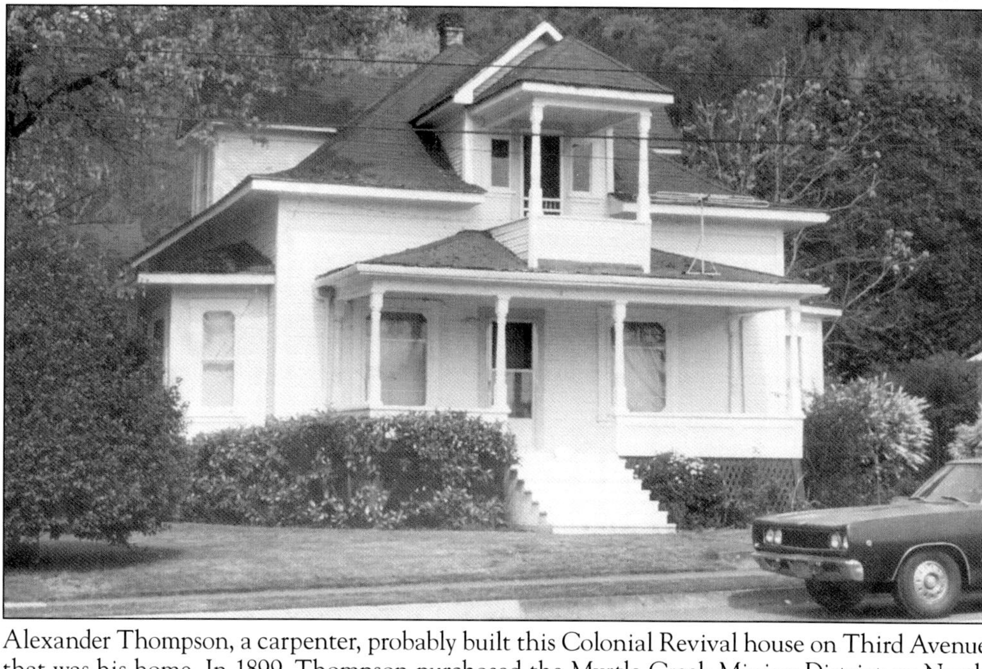

Alexander Thompson, a carpenter, probably built this Colonial Revival house on Third Avenue that was his home. In 1899, Thompson purchased the Myrtle Creek Mining District on North Myrtle Creek. The purchase included all ownership and water rights of Verwood, Mulkey, Yocum, and Catching. The district was a major center of gold mining activity in the 1890s. (Courtesy of Florence Chaney.)

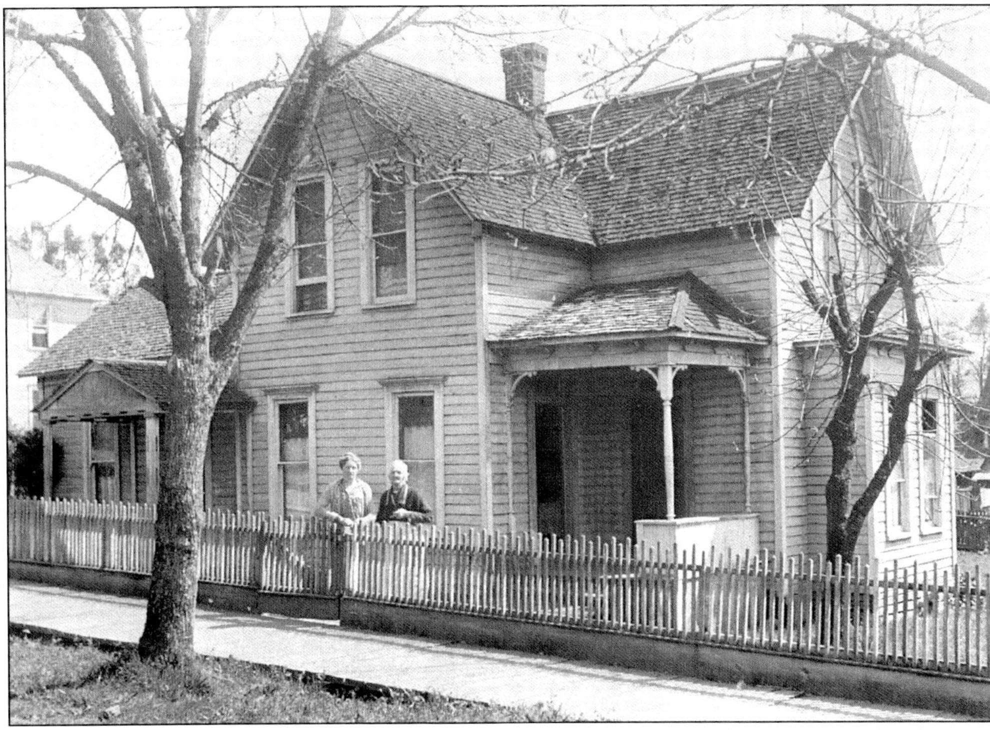

Dr. B.F. Fallin, Sarah Jane Fallin, and their daughter Susie traveled over the Oregon Trail and arrived when Susie was three years old. Doctor Fallin was a physician, and his wooden medicine chest can be seen at the Douglas County Museum. The house above was built for them at the corner of Third Avenue and Oak Street around 1882. Doctor Fallin's office was in the bay that protrudes by the front porch. The woman to the left above is Susie Fallin Dailey (Mrs. Mervin Dailey), who was well known for her 73 years of attendance and service to the Christian Church. The other person is unidentified. Below, Doctor Fallin and a companion pose in his buggy in front of the Bilger Creek home of daughter Nancy and son-in-law Samuel Dailey. (Both, courtesy of Joyce Stimpson.)

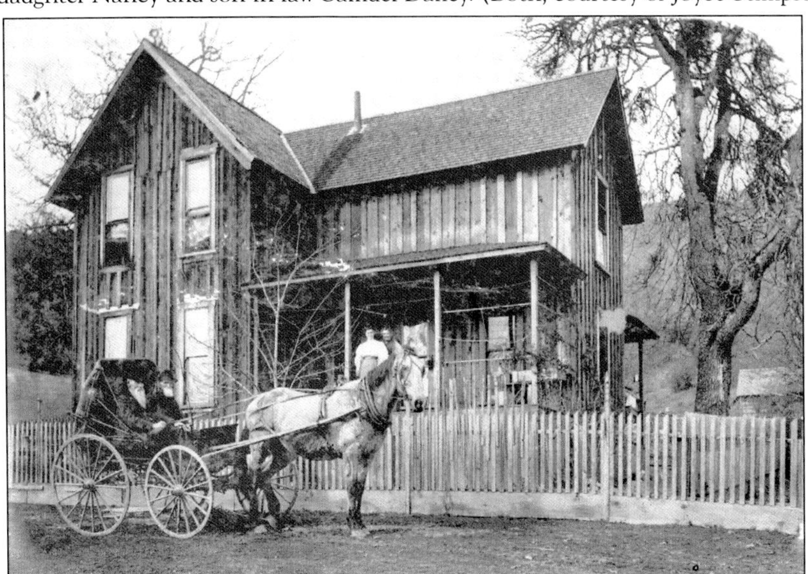

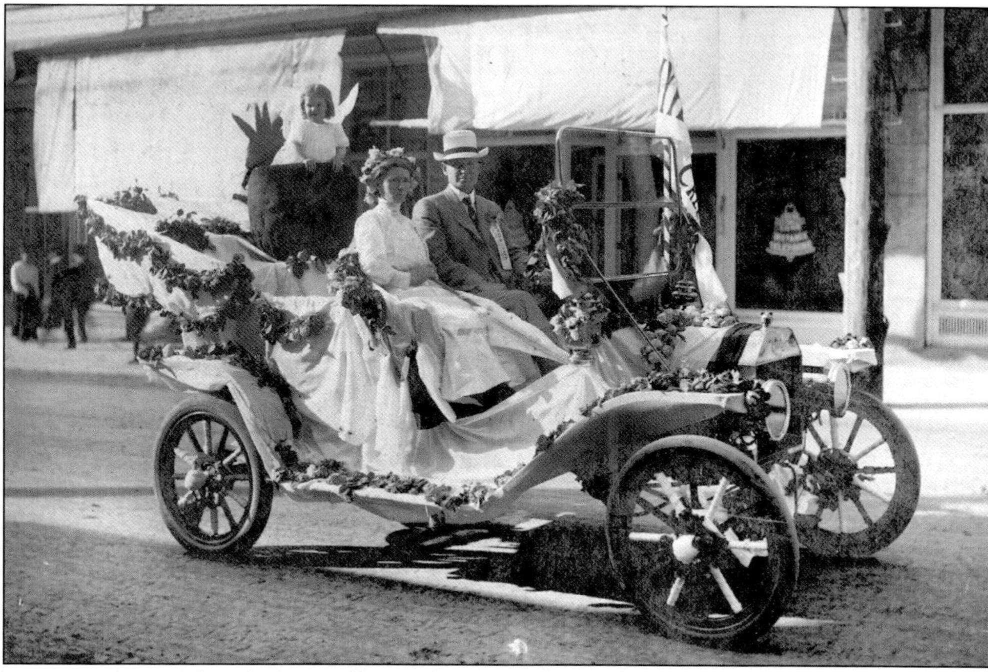

Passing in front of the telephone office in the Fate building on Second Avenue is a gaily-decorated Model T. This touring car was made by Ford from 1910 to 1912, which helps date this photograph. The festival queen and her escort are parading a young girl perched in a replica strawberry. She is probably the Strawberry Princess of the spring festival. (Courtesy of Canyonville Pioneer-Indian Museum.)

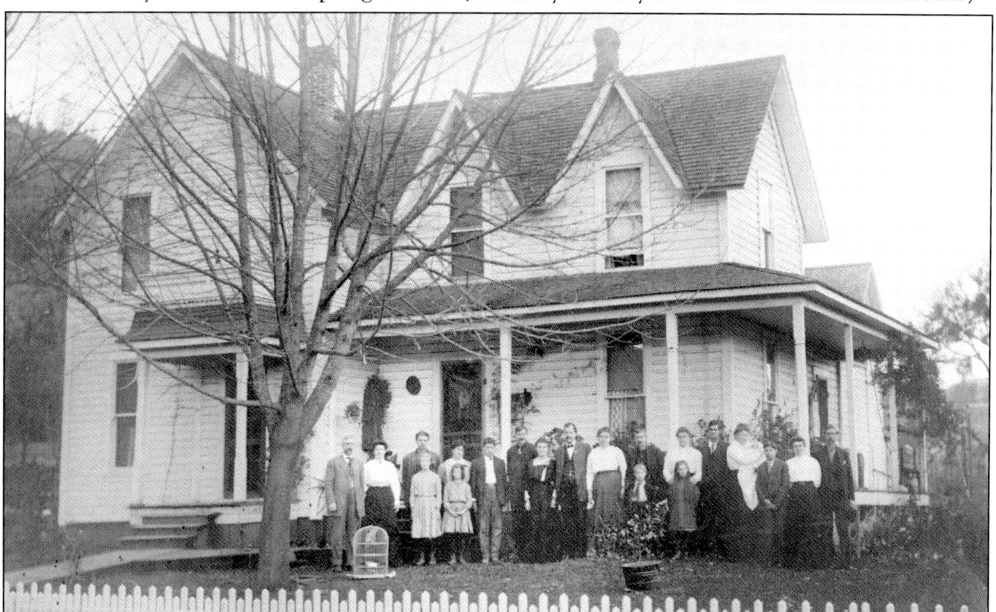

The extended family of Remick Fate gathered for a 1911 New Year's celebration at his home on Hall Street. Fate planted the landmark Redwood tree at the corner of Hall Street and Second Avenue. Besides owning and farming 500 acres near town, Fate owned the large commercial building on Second Avenue next to Citizens State Bank. James Rice lived in this house until 1909.

Five

Roads and Rails

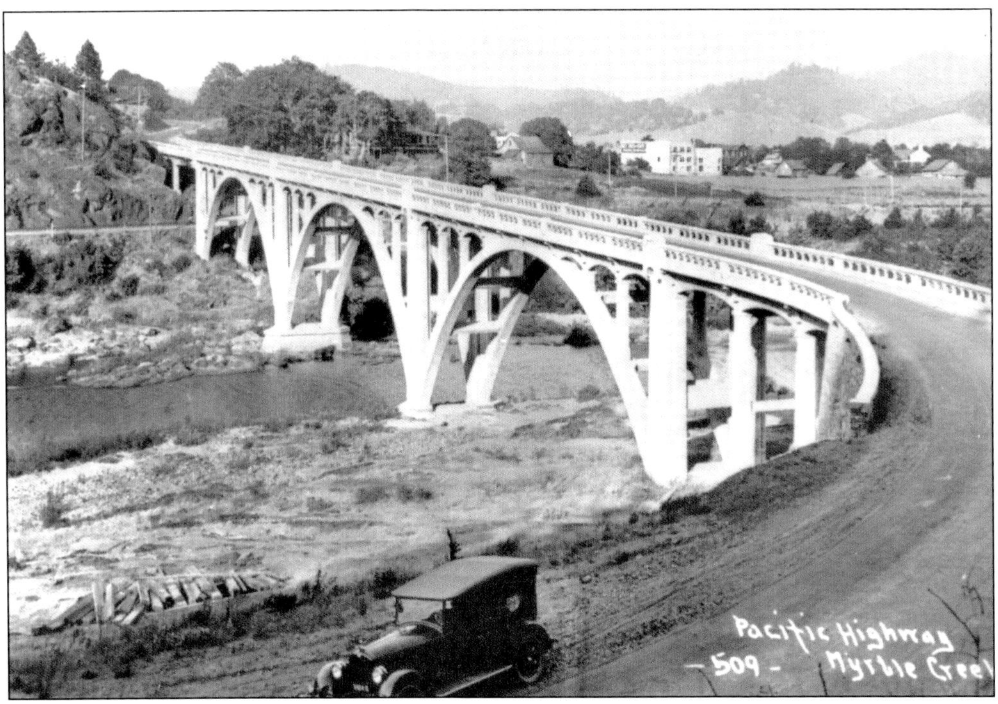

The new entrance to Myrtle Creek is completed at last! Revealed in this 1922 photograph, the beauty and grandeur of the new bridge is a wonder to behold. Most bridges are aligned with the road so travelers cannot see what they look like, but the 90-degree approach to the South Umpqua Bridge offers Pacific Highway 99 travelers a full view of its artistic architecture. (Courtesy of City of Myrtle Creek and Bill Whisenant.)

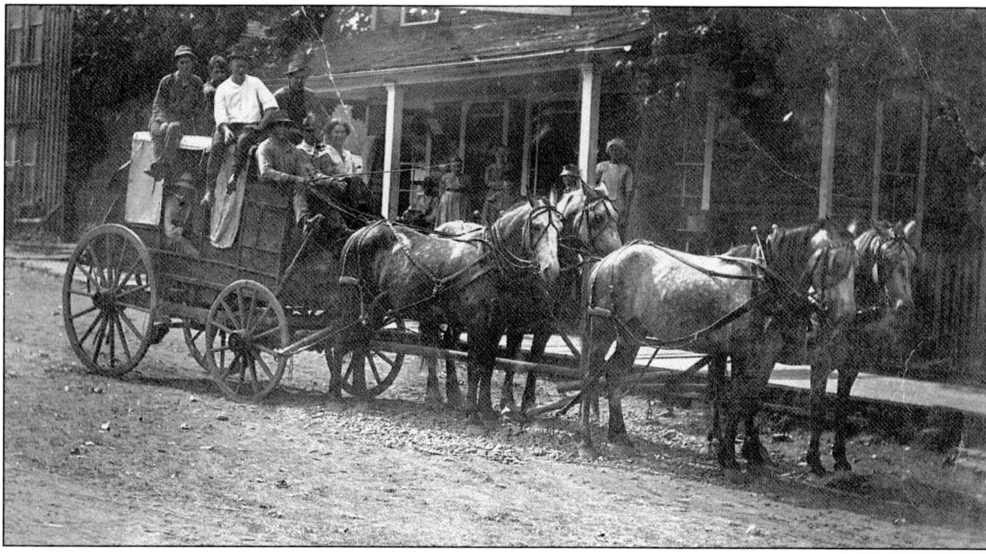

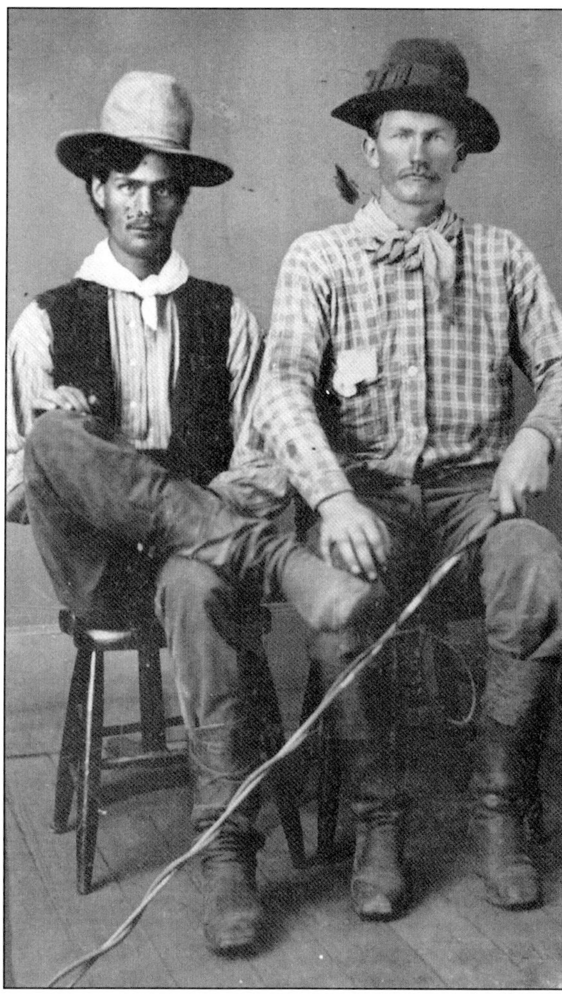

For the first 30 years of Myrtle Creek's existence, the only road into town was the Scott-Applegate Trail. During those years, it had undergone improvements and name changes, but the route remained the same. By 1860, the trail had been improved by the military and was dubbed the Hooker Military Road. Stage service began in 1859, providing Myrtle Creek with its first public transportation system. Tom Burnett, holding the whip at left, drove his first four-horse stage coach from Round Prairie to Roseburg at age 17 and continued as a longtime driver for the Oregon and California Stage Company.

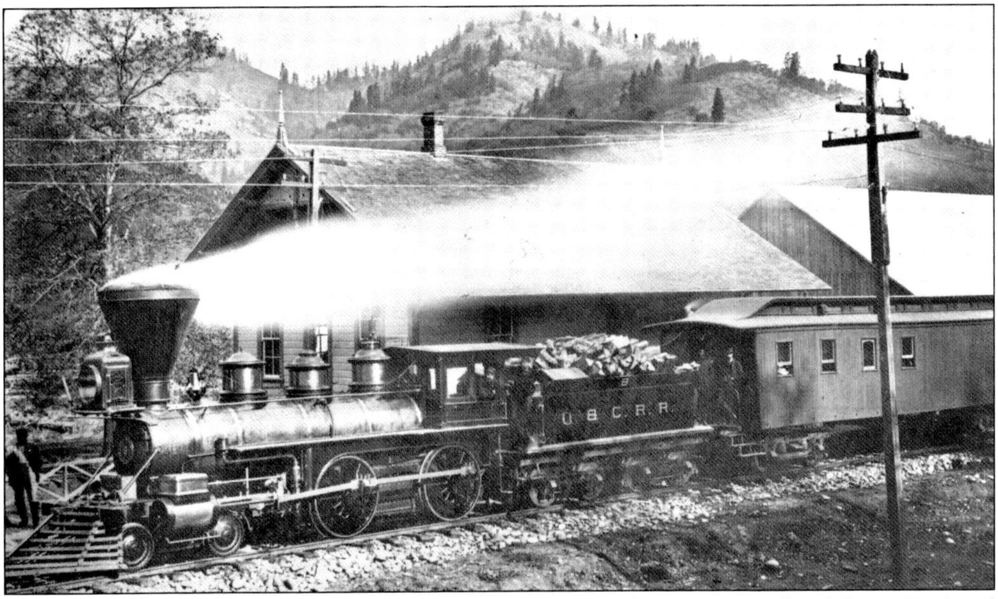

The first train to Myrtle Creek arrived on August 1, 1882. The Oregon and California Railroad was critical to the development of Myrtle Creek. Besides creating jobs for woodcutters, station agents, and telegraph operators, the railroad facilitated the shipment of farm crops, lumber, and people. According to the signs at the depot, the Myrtle Creek station was 550.4 rail miles from San Francisco at an elevation of 637 feet.

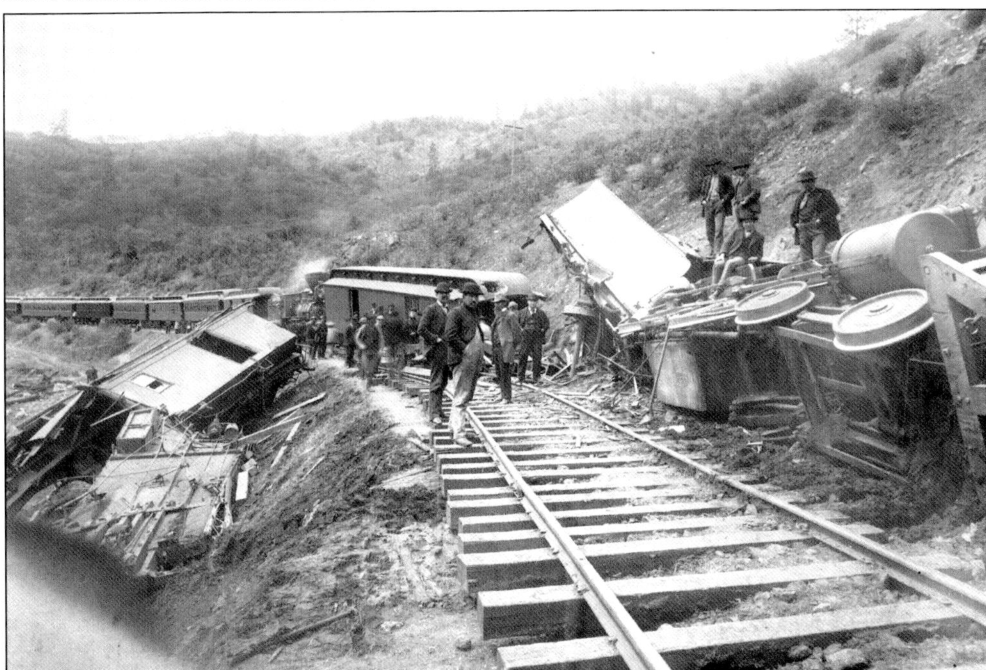

On April 27, 1892, a southbound train was derailed by railroad ties left on the tracks a mile north of Myrtle Creek. Unfortunately, no local newspapers survive to provide details, but a Eugene newspaper claimed the derailment was an attempted robbery of a gold shipment to be picked up in Myrtle Creek. If that was the case, why was this train derailed before it reached the depot?

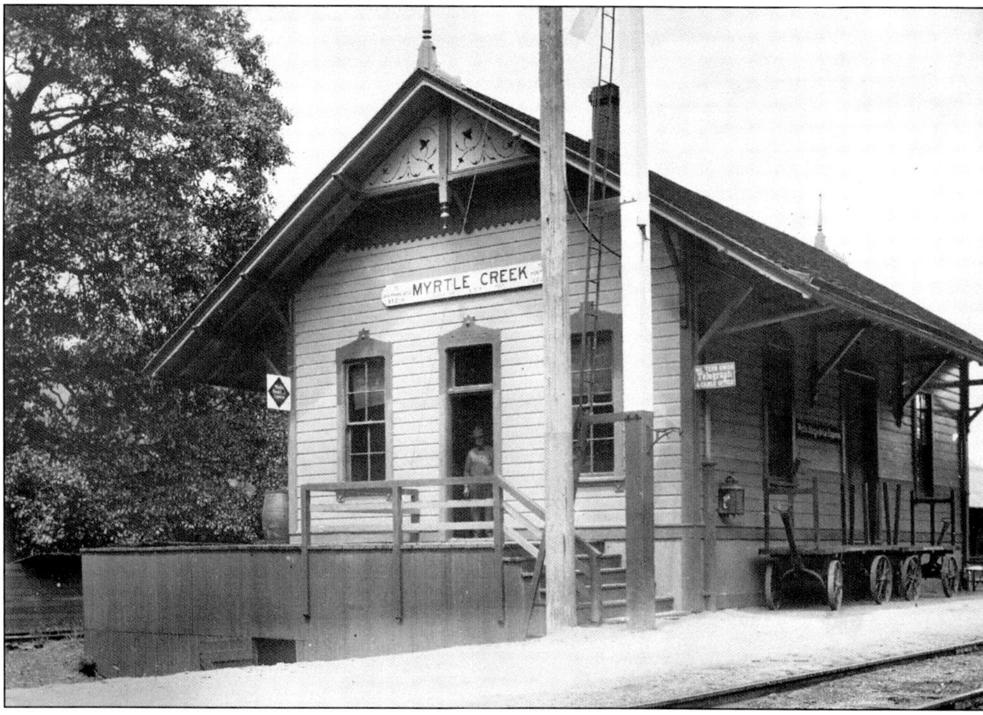

Fifty-three years after the Myrtle Creek depot was established, a 1935 newspaper article reported on the citizens' continued unhappiness about how far from town the O&C Railroad had placed the station. This remoteness may have been a factor in a January 1897 robbery, when transient Frank Hoyt pistol-whipped Agent Hoopengarner and fled with a small sum of money. Recognized by Hoopengarner, Hoyt was arrested the next day in Roseburg.

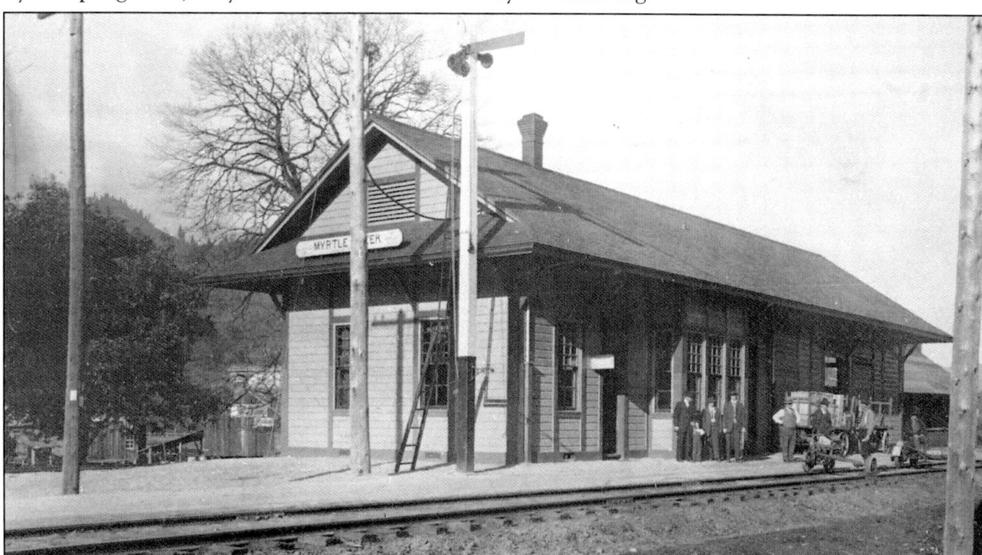

Around 1915, the Southern Pacific Railroad built a new train station in Myrtle Creek. The new station provided expanded cargo storage and indoor accommodations for the growing passenger base. Some residents used the railroad to commute to work in Roseburg. In 1923, a roundtrip ticket cost 52¢.

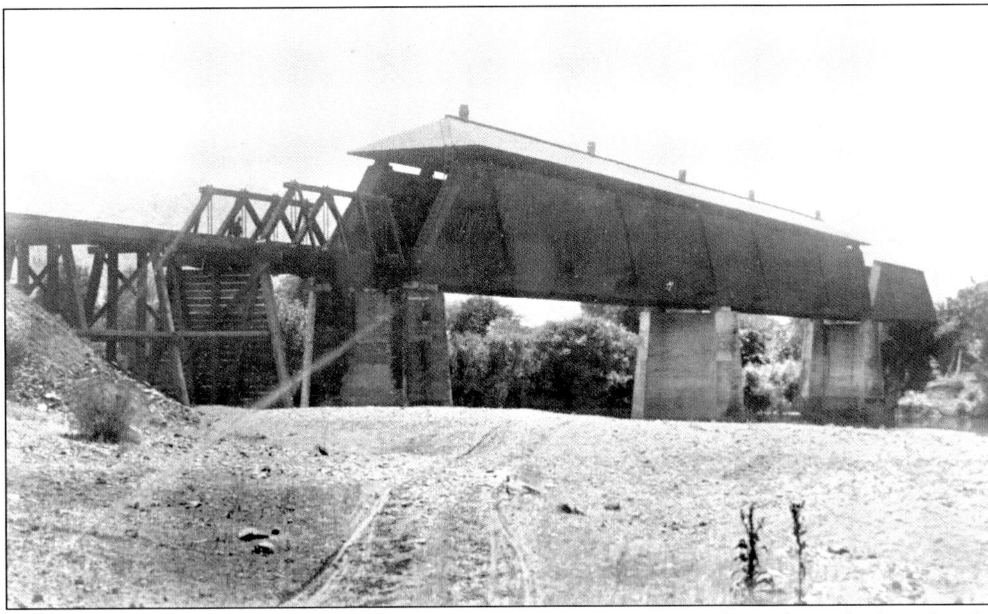

This covered trestle at Myrtle Creek was built to replace the original trestle washed out by the 1890 flood. The new trestle was built in haste because the flood also caused an expansive slide over the tracks in the Cow Creek Canyon, and support trains from the north were needed to help relocate more than three miles of track. This wood trestle was replaced in 1906 with the current steel span.

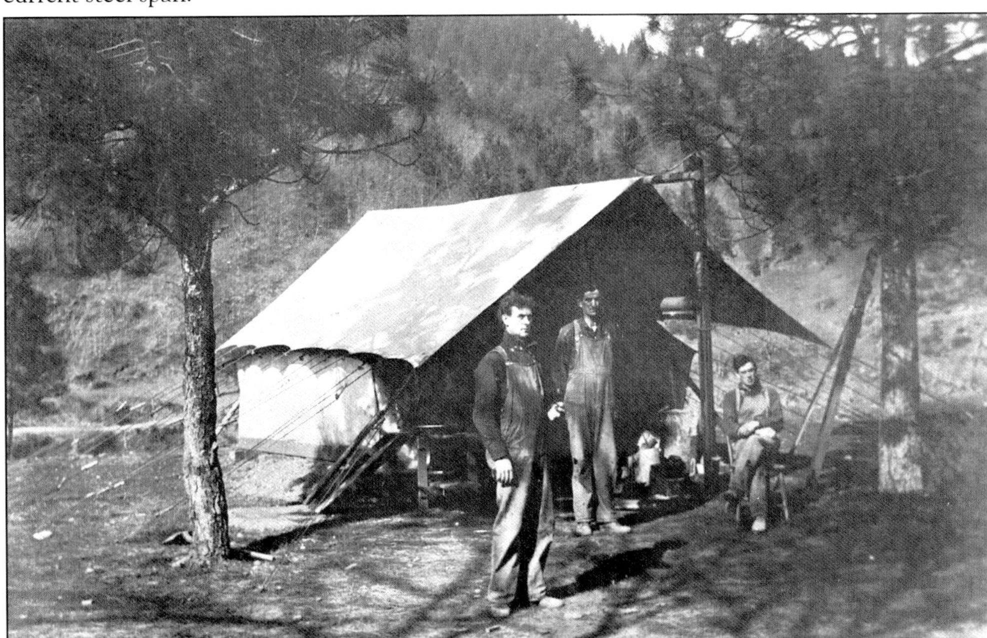

This undated photograph shows a railroad work crew camp below the tracks just north of town in Hadley Flat. Because of the number of trains in use and the number of people who relied on them, numerous work crews were necessary to maintain the tracks and provide dependable service. These work crews also provided a source of jobs for local laborers.

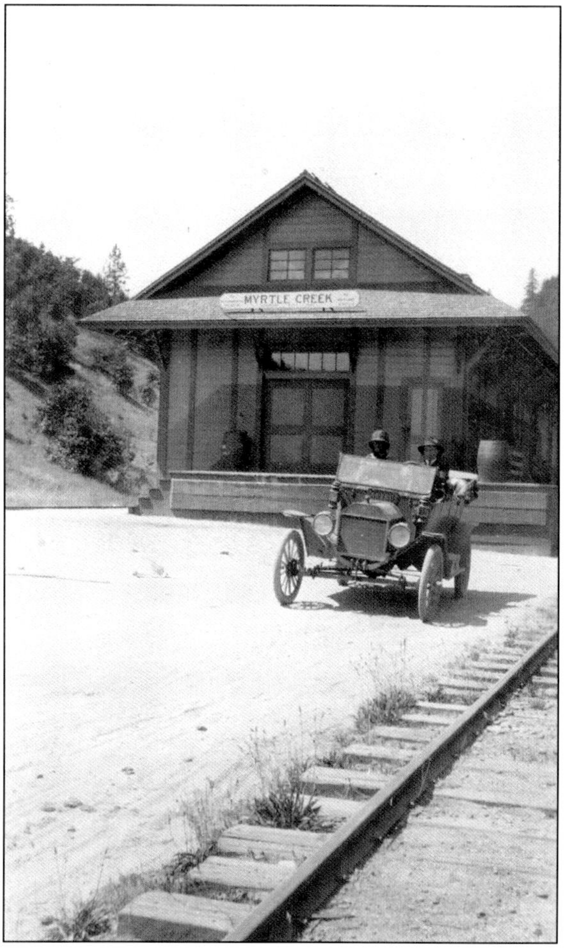

A couple in a late-1913 Model T Ford touring car poses at the train depot. From the time the railroad arrived in 1882 until the depot was closed around 1955, the area around the depot was a hub of activity with a hotel, packinghouses, and mills. In the 1920s, there was a boardwalk that went from the depot to town, and from town out to the school on Division Street.

In this 1914 photograph, Earl Sumner stands with his Indian motorcycle. If he had been on the road very long, he probably needed a bath, as paved roads in the vicinity of Myrtle Creek were still about eight years away. The nearest pavement was on the streets of Roseburg, nearly 20 miles to the north. (Courtesy of Dora Dyer.)

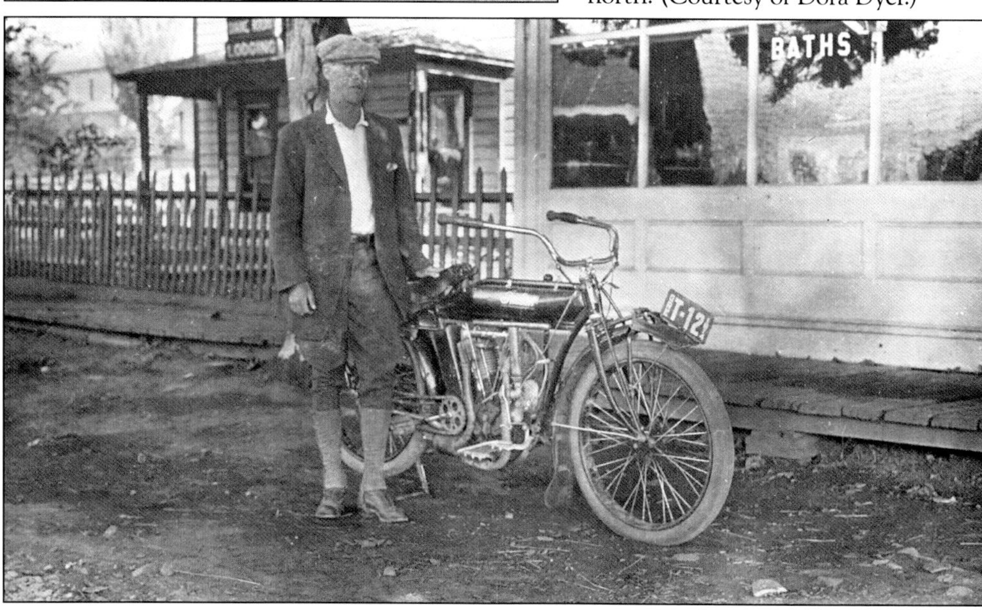

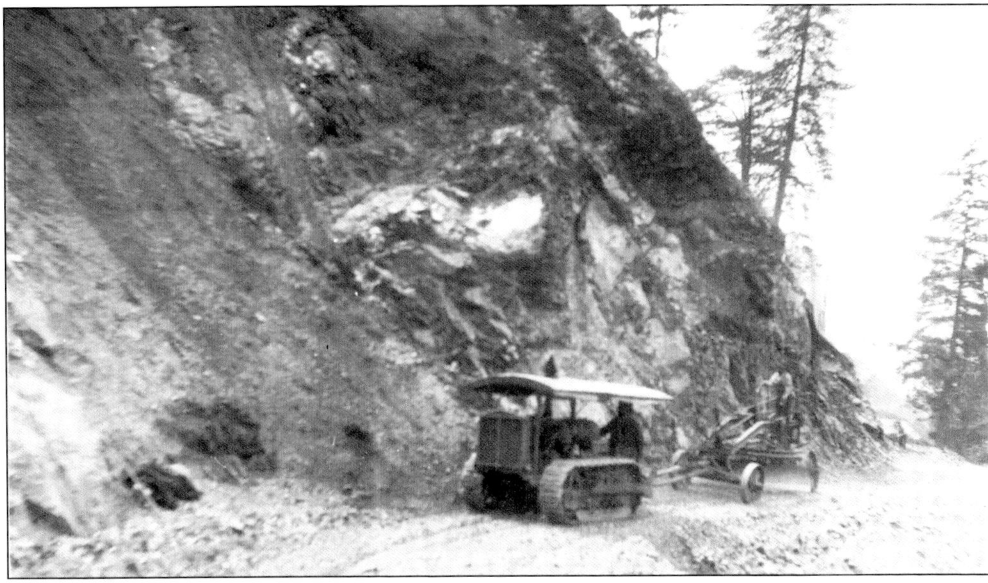

These 1918 photographs taken by Ellen Rice show construction in progress on Pacific Highway 99. The tractor above is pulling a grader at Serpentine Point, about one mile north of Myrtle Creek. A little farther north, the crew (below) is using horses and fresnos to scrape away soil to clear the roadbed. When this federally funded project finished the highway to Myrtle Creek in 1922, the increase in traffic through the heart of town had an immediate and significant impact on the local economy. Tourism became the new catchword, and local businesses thrived on it. (Both courtesy of Bill Rice.)

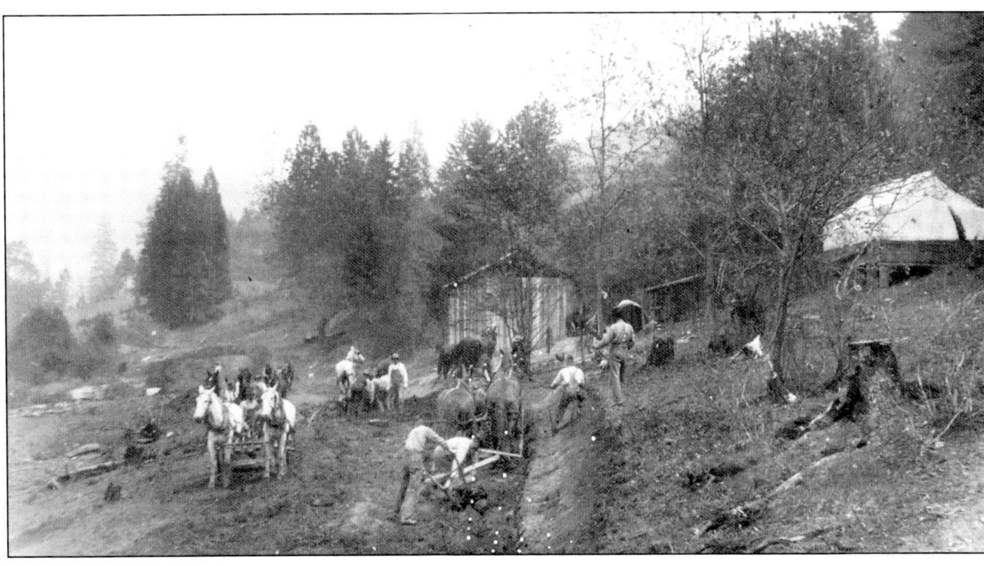

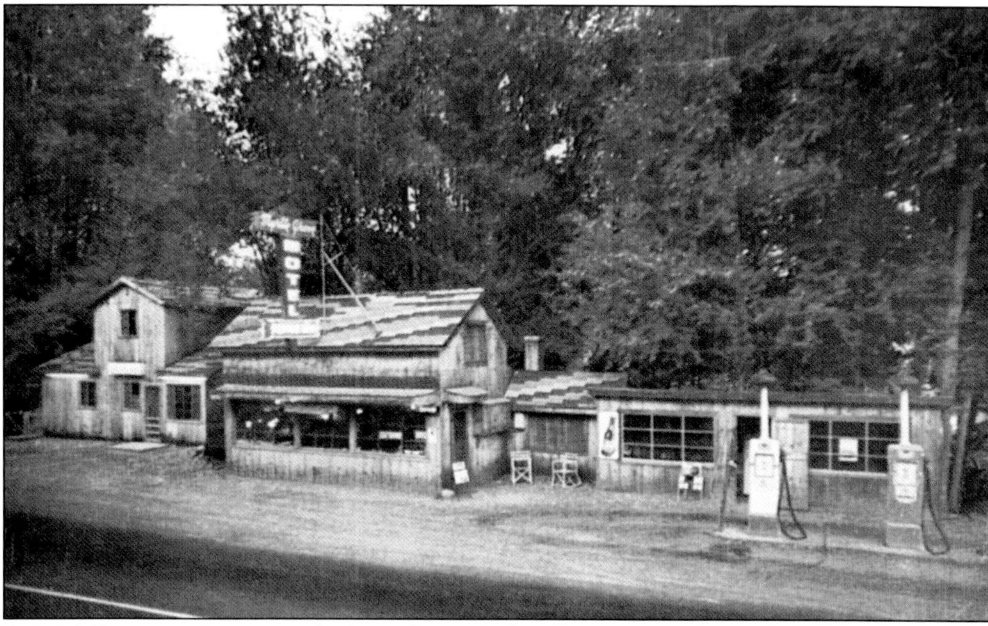

Beautifully situated in a stand of Myrtle trees, the Myrtle Grove Motel, coffee shop, and gas station served Highway 99 travelers until 1965. Located about six miles north of Myrtle Creek, it featured handcrafted Myrtlewood items popular with tourists, and cabins popular with fishermen. The loss of the cabins in the 1964 flood and the loss of traffic to the new Interstate 5 led to its demise. (Courtesy of Dennis Weber.)

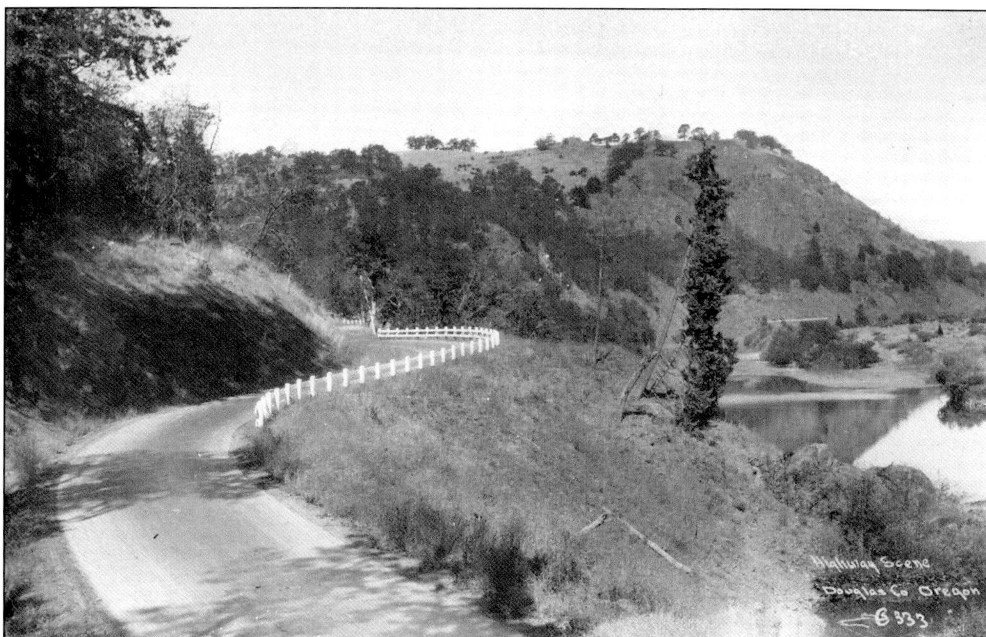

This scenic view of the South Umpqua River along Pacific Highway 99 looks north from the highway about six miles north of Myrtle Creek. Just around the corner ahead, the Myrtle Grove Motel, which was a popular destination for fishermen and diners, sat on its perch and looked down on the river.

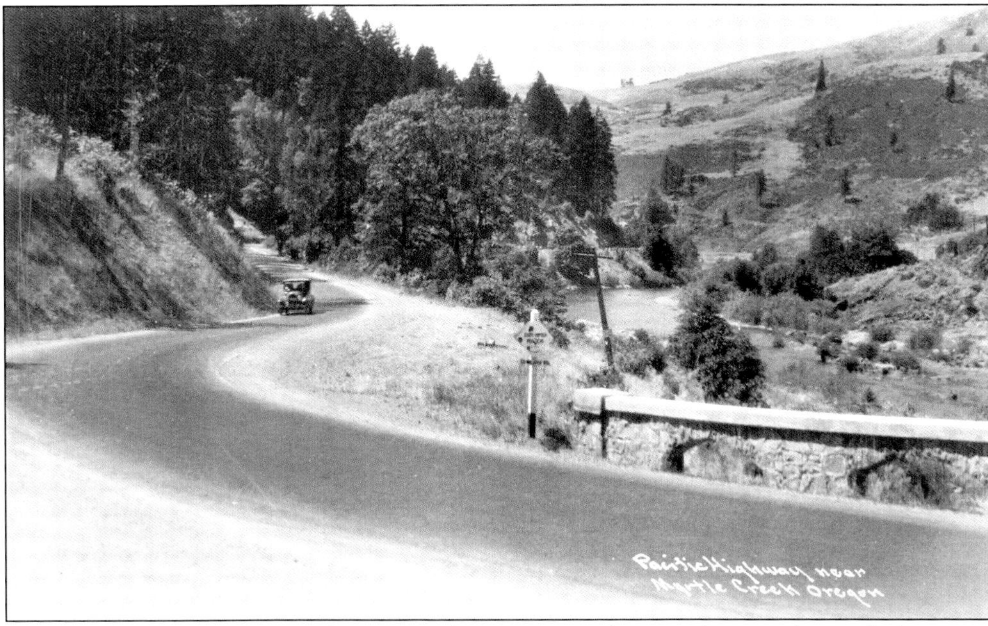

Shortly after the completion of Pacific Highway 99, an unknown photographer took this scenic shot of the South Umpqua River that includes a rare, late 1910s or early 1920s Flint automobile approaching the new bridge entrance to the bustling city of Myrtle Creek.

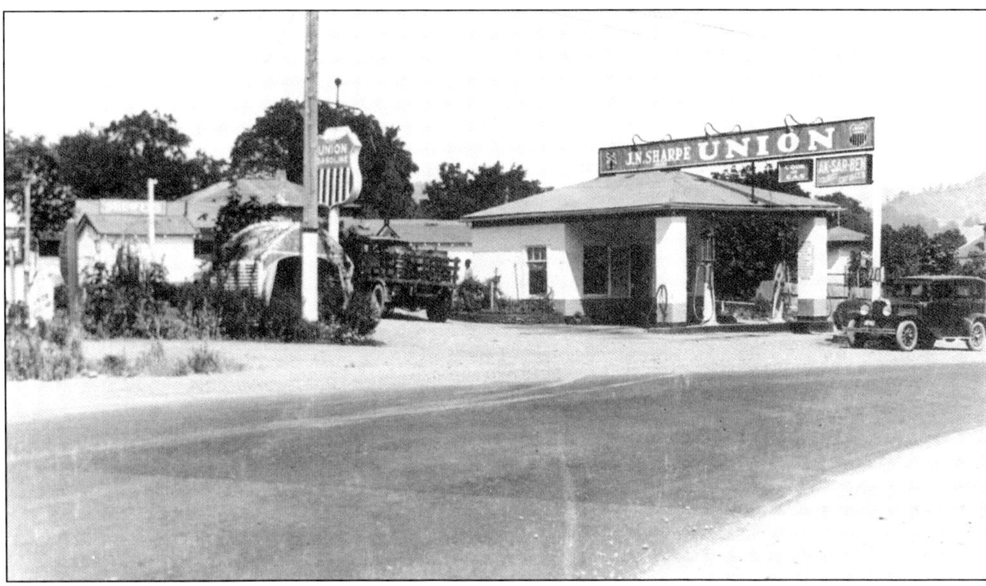

J.N. Sharpe's business, located at the top of Main Street where Doctor Goodwin's Optometry Office now stands, was known as the Ak Sar Ben (Nebraska spelled backwards) Tourist Cabins and Union Gas Station. This was reportedly the first establishment on Pacific Highway 99 to combine these features. This combination of motor and hotel services resulted in the coining of a new word: motel.

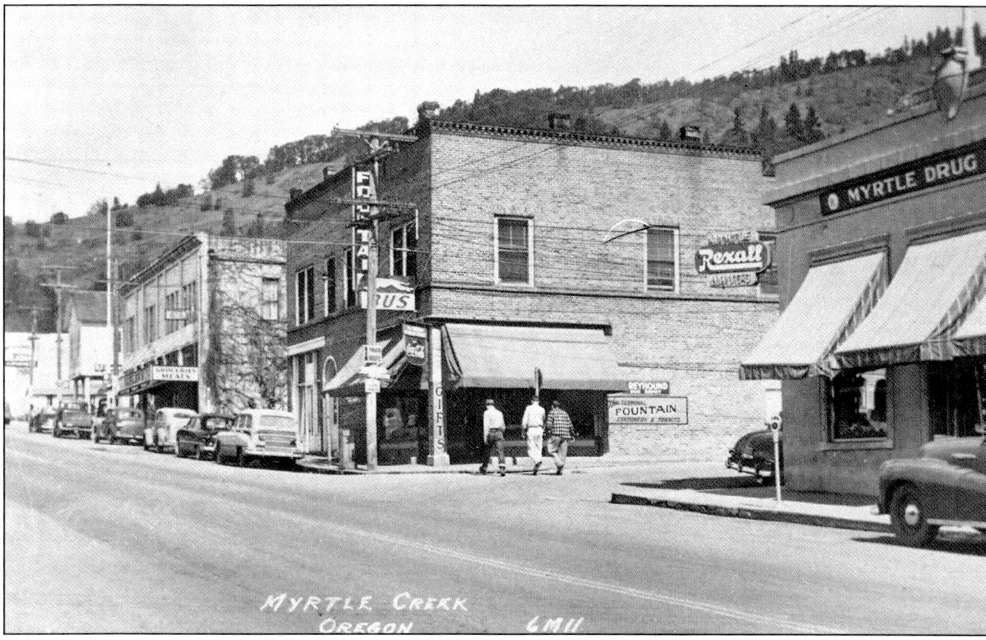

This c. 1950 street scene of Myrtle Creek's Main Street shows the Myrtle Creek Rexall Drugstore, and across Second Avenue is the Weaver building, home of the Terminal Fountain and Greyhound bus depot. The Myrtle Creek Hospital was on the second floor of the Weaver building, and up the street, the sign for Marion Kusler's market in the IOOF building advertises groceries and meat.

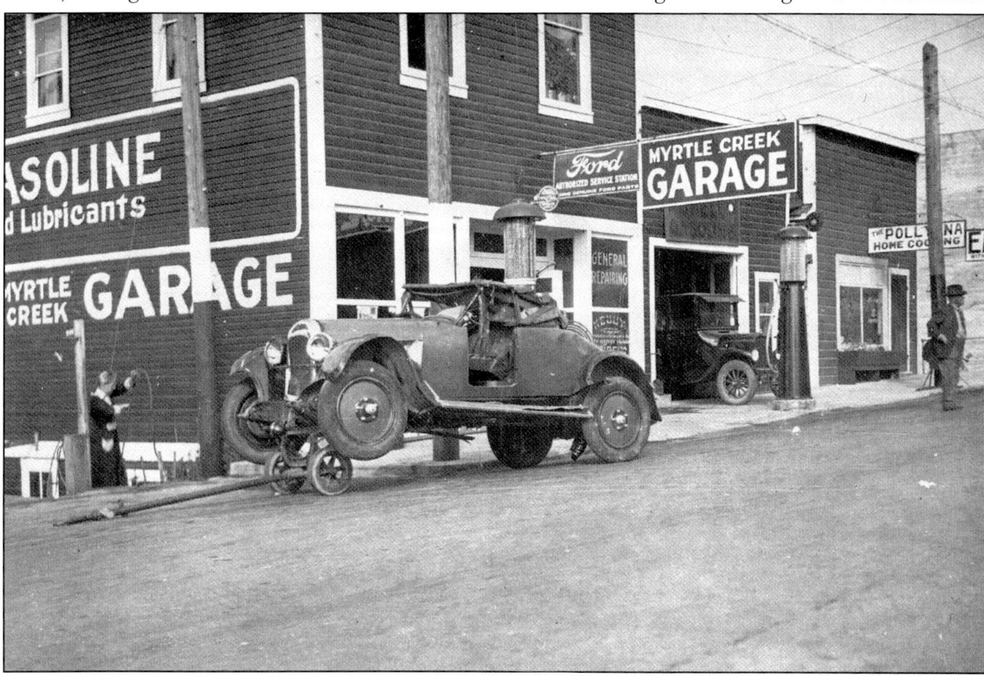

A wrecked 1926 Nash Roadster sits on a dolly in front of H.C. Shirtcliff's Myrtle Creek Garage with a 1925–1926 Model T Ford in the garage door. Established in 1921, the garage was owned by Henry Shirtcliff and was operated by his son Ted. This building burned and was replaced with the current building in 1933. Watering flowers beside the building is Ted's wife, Mae Shirtcliff.

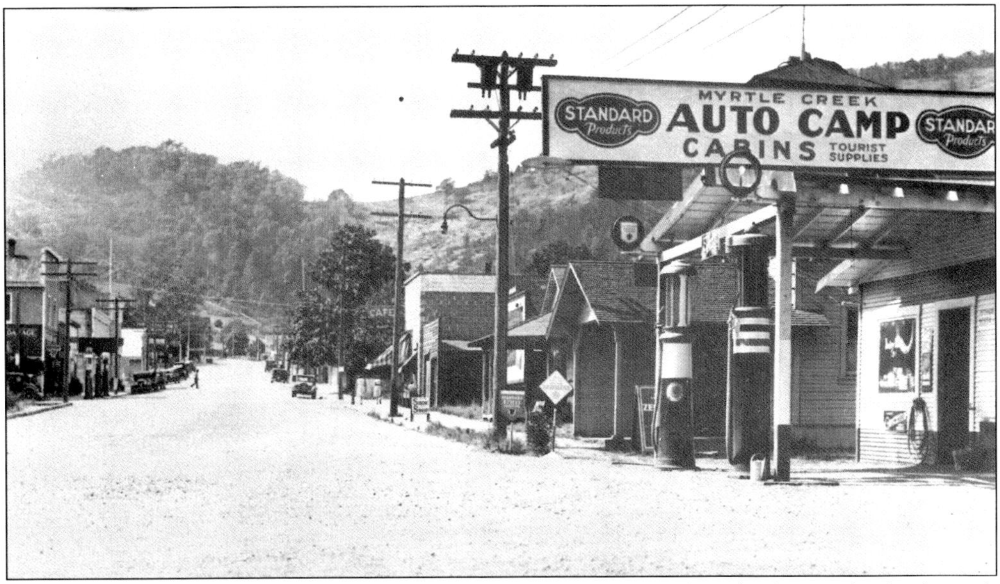

Pacific Highway 99 ran from Tijuana, Mexico, to Vancouver, British Columbia, and right through the heart of Myrtle Creek, benefiting local businesses with a steady stream of tourists and travelers. Places such as the Myrtle Creek Auto Camp and Standard station emerged to service this increased traffic. The Myrtle Creek Garage (far left) and the Ak Sar Ben Tourist Cabins at the north end of town were also established in the 1920s.

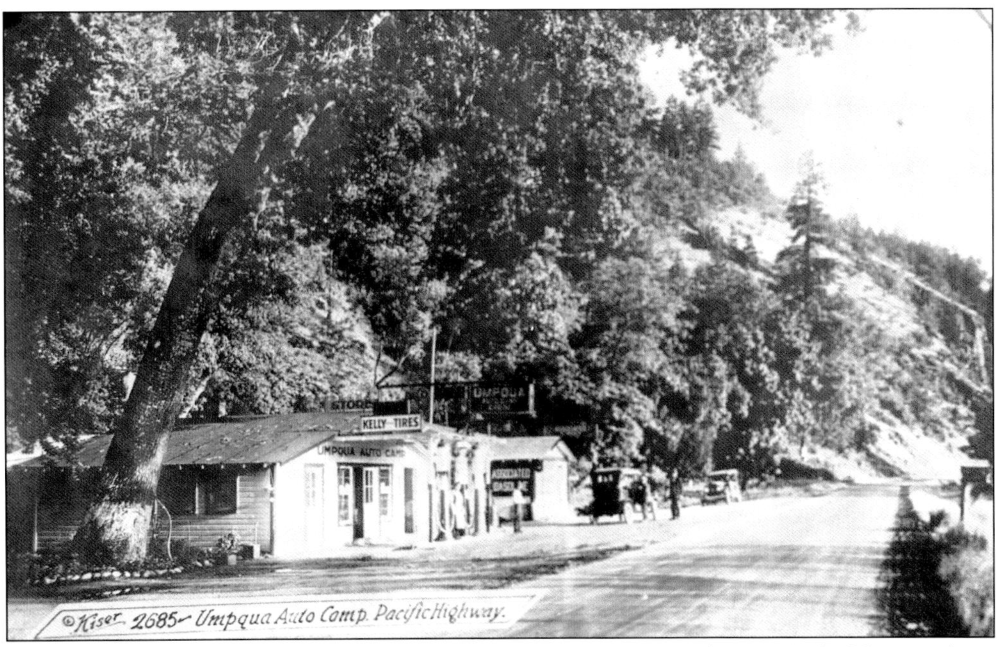

South of Myrtle Creek, on the curve south of what is now Interstate 5 exit 103, the Umpqua Auto Camp sat on the east side of the road across from the South Umpqua River. Featuring Kelly tires and three glass-topped, gravity-fed gas pumps, this popular tourist stop is captured in this early 1920s photograph.

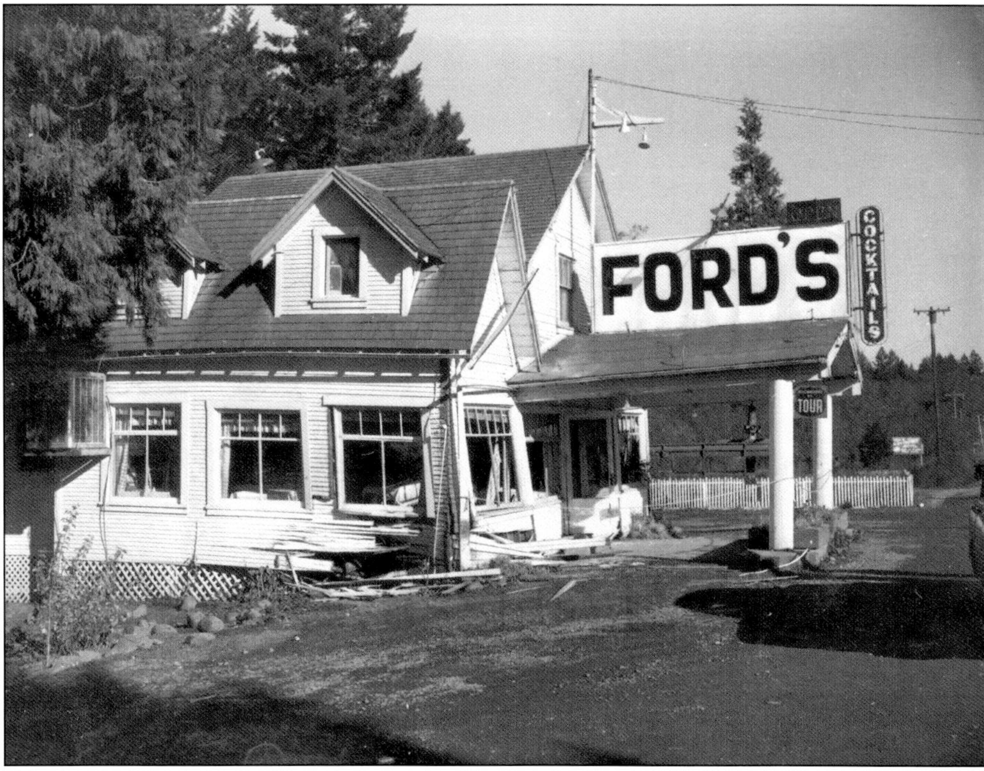

Because of its famous fried chicken and petting zoo (below), Ford's Restaurant was a popular stop for Pacific Highway 99 travelers in the 1940s and 1950s. Located between Myrtle Creek and Canyonville at the southeast corner of the South Umpqua River Bridge at Gazley Road, it was razed in the late 1950s soon after the photograph above was taken. The site now lies under Interstate 5. (Both, courtesy of Betty Hanna.)

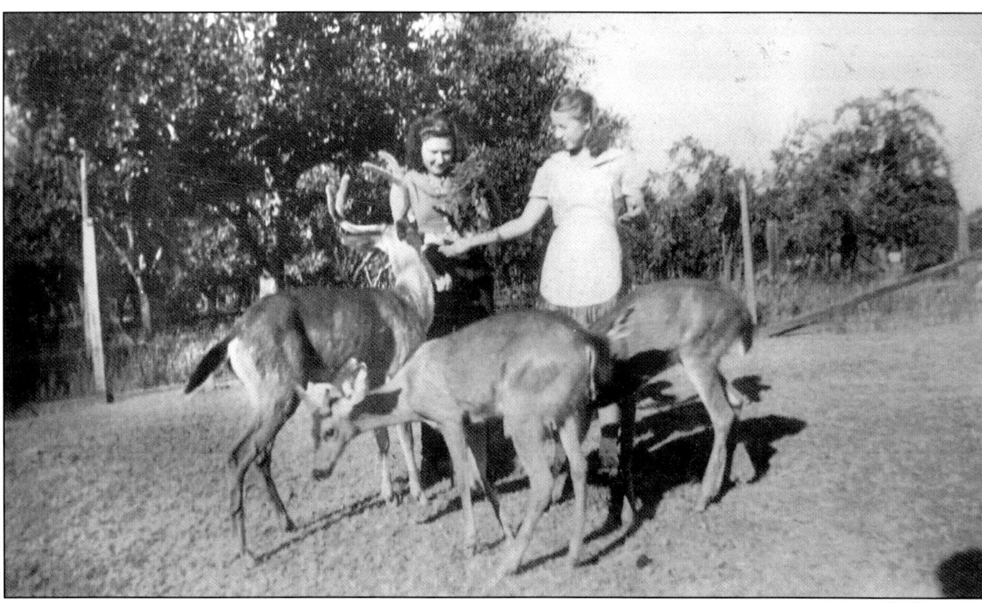

Six

Roaring 1920s and Depressed 1930s

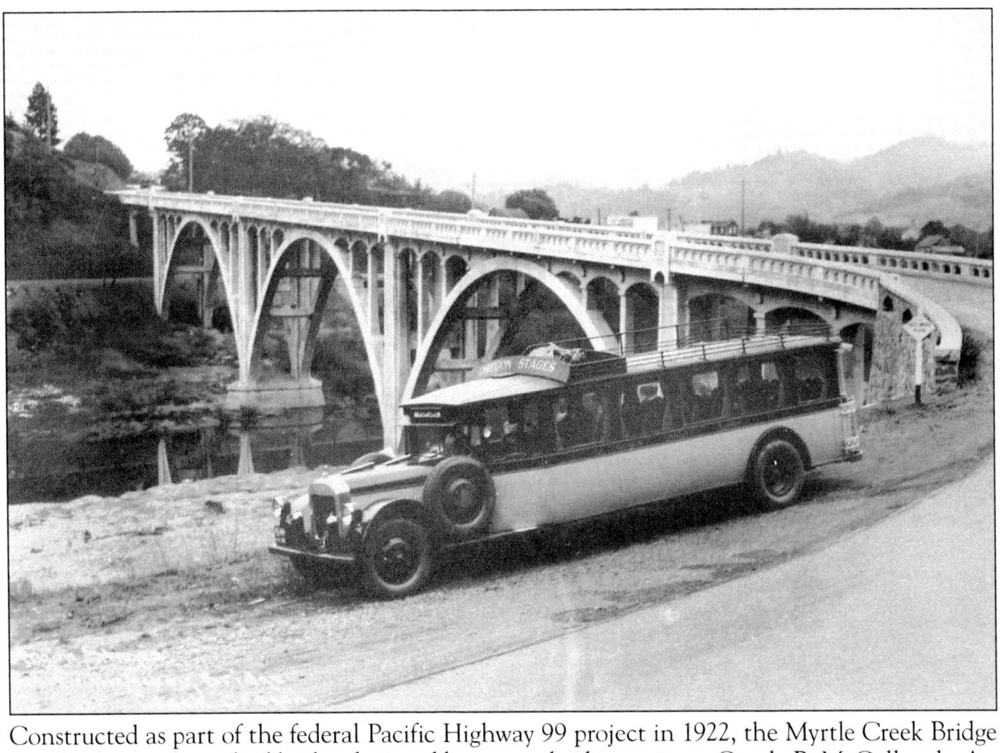

Constructed as part of the federal Pacific Highway 99 project in 1922, the Myrtle Creek Bridge was the first multi-arched bridge designed by master bridge engineer Conde B. McCullough. An engineering professor at Oregon State College, he was the bridge engineer for the state of Oregon. Using this Myrtle Creek Bridge design, McCullough later engineered the famous bridges on the Roosevelt Highway, now known as Highway 101.

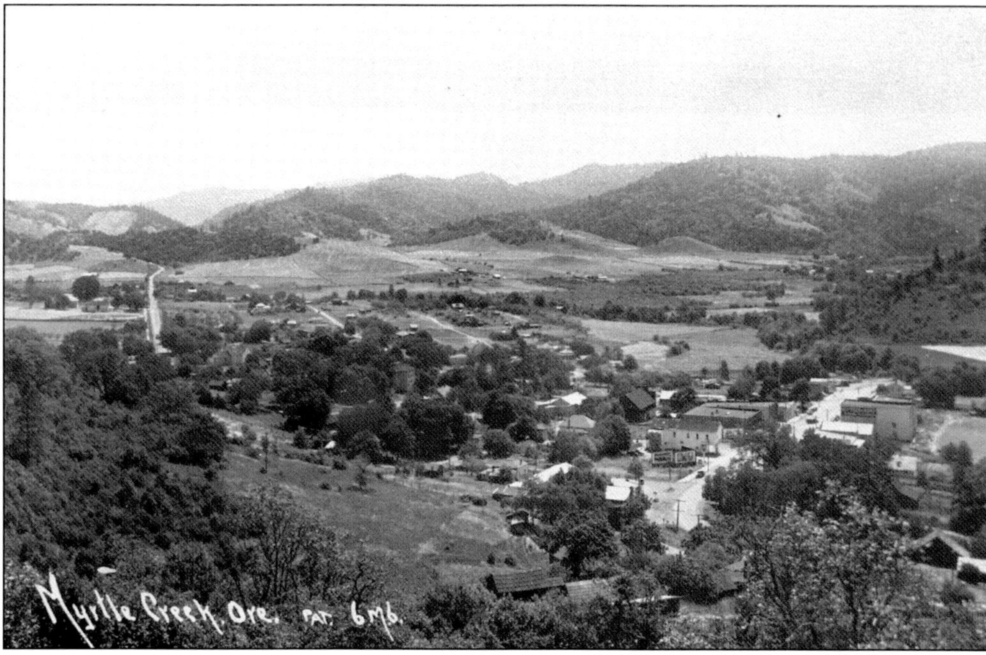

Taken from the hill above Rooster Rock, this c. 1935 photograph depicts Myrtle Creek as a quiet little village nestled in a beautiful valley. The Rice building is now home to the Reynolds and Adams store, and where the Myrtle Inn and Ignoffo's Italian Bistro now stand, there is a large sign advertising Old Gold cigarettes.

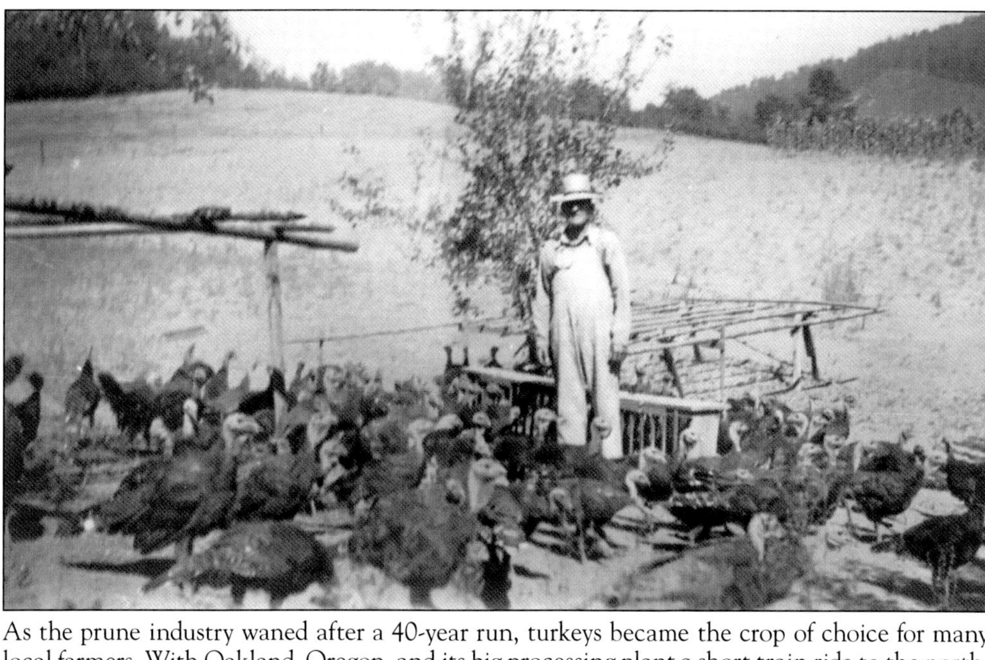

As the prune industry waned after a 40-year run, turkeys became the crop of choice for many local farmers. With Oakland, Oregon, and its big processing plant a short train ride to the north, there was a ready market available. Many prune orchards were converted to sunflower fields to feed the expanding turkey population. Here, Wally Brownson poses with his flock of turkeys. (Courtesy of Martha Brownson.)

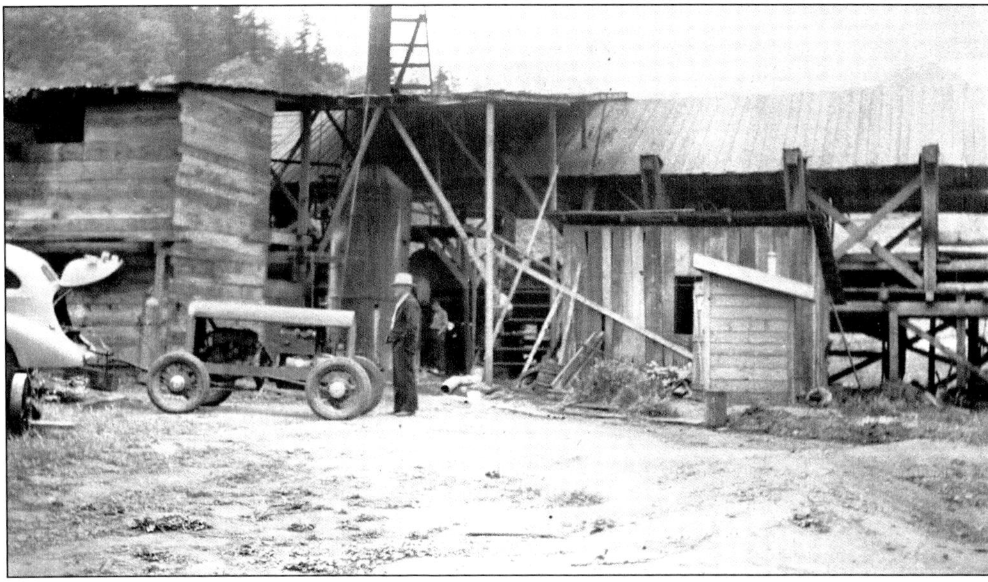

The back of a 1937 Hudson Terraplane can be seen at left in this 1937 photograph of Daly's steam-powered sawmill. Located across the tracks behind the train station, this mill was typical of the many small "gyppo" sawmills spread across the timber country of southern Oregon in the post-Depression days.

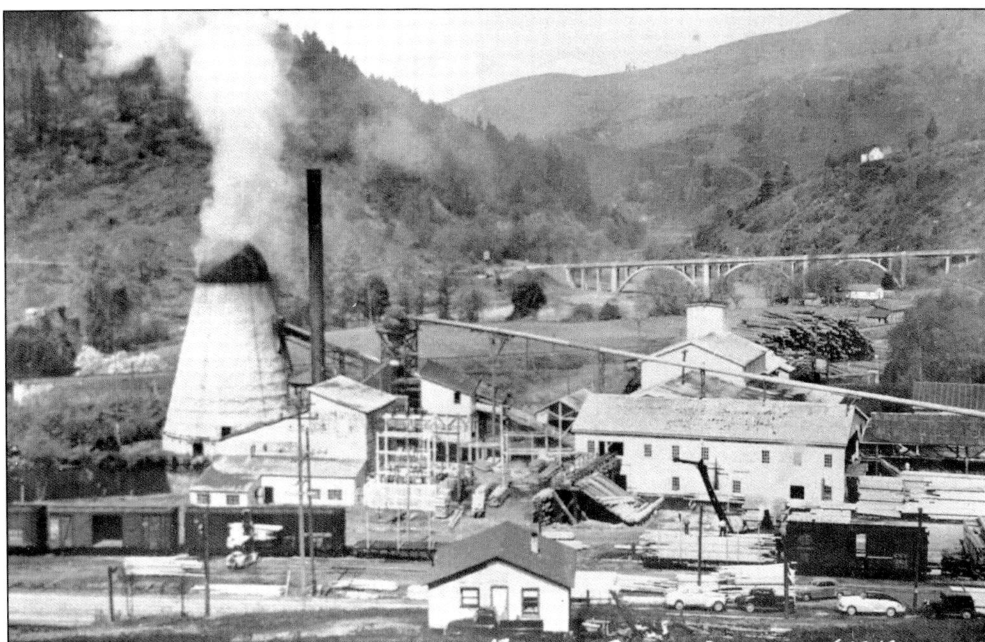

Now icons of the past, wigwam burners were the prominent feature of post–World War II lumber mills. A conveyor carried wood scraps and sawdust up a ramp to an opening near the top and dumped the waste into a roaring fire. The demise of the wigwam burner began when the development of chipboard created a market for wood chips. Within a few years, wigwam burners were no longer in use.

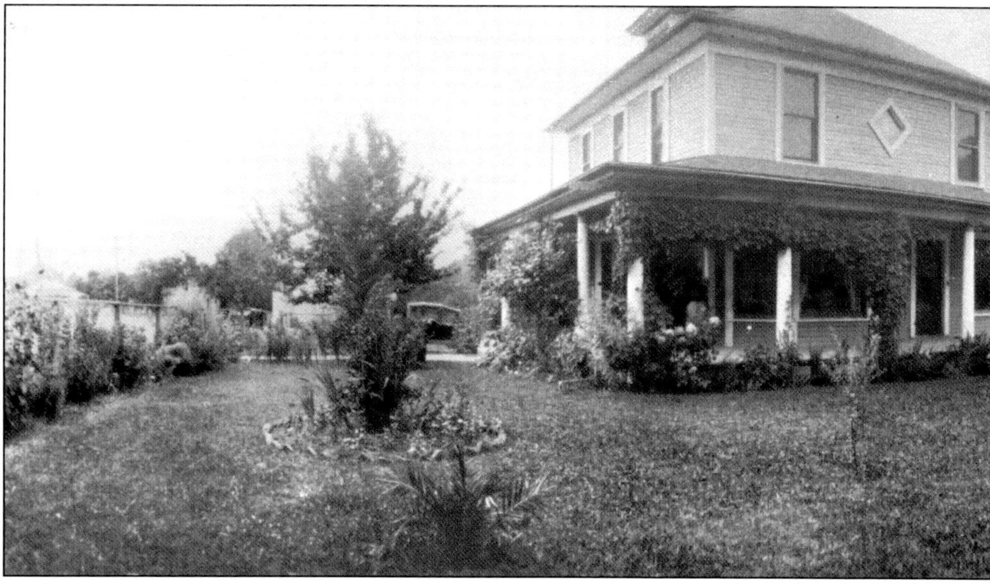

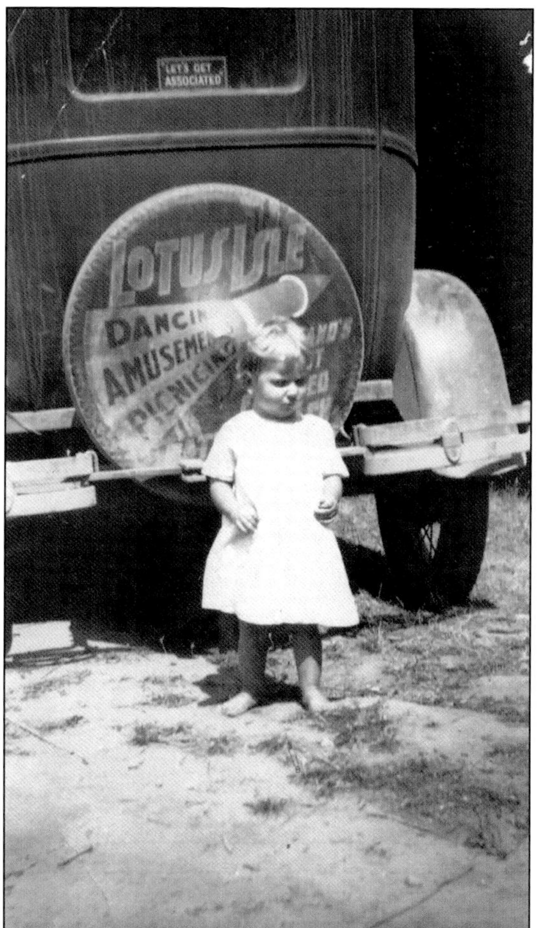

When this spacious house was new, Myrtle Creek sported a population of about 400 in 1909, and Mayor Samuel S. Johns and his wife, Mary, lived here. According to the 1910 census, they were taking in boarders, as was later owner "Pop" Jackson. Located on First Avenue between Oak and Broadway, it was torn down and replaced in 1968 by Hall's Western Auto. (Courtesy of Blanche Newton and Steve Hall.)

This cute young lady standing barefoot behind a 1927–1928 Model A Ford is Dora Smith Dyer, who would live to know a mode of travel far different than what her great-grandfather knew. He was Anderson "Poker" Smith, and he crossed the continent on the first wagon train to the Oregon Territory, arriving in 1843 with the Applegates. (Courtesy of Dora Smith Dyer.)

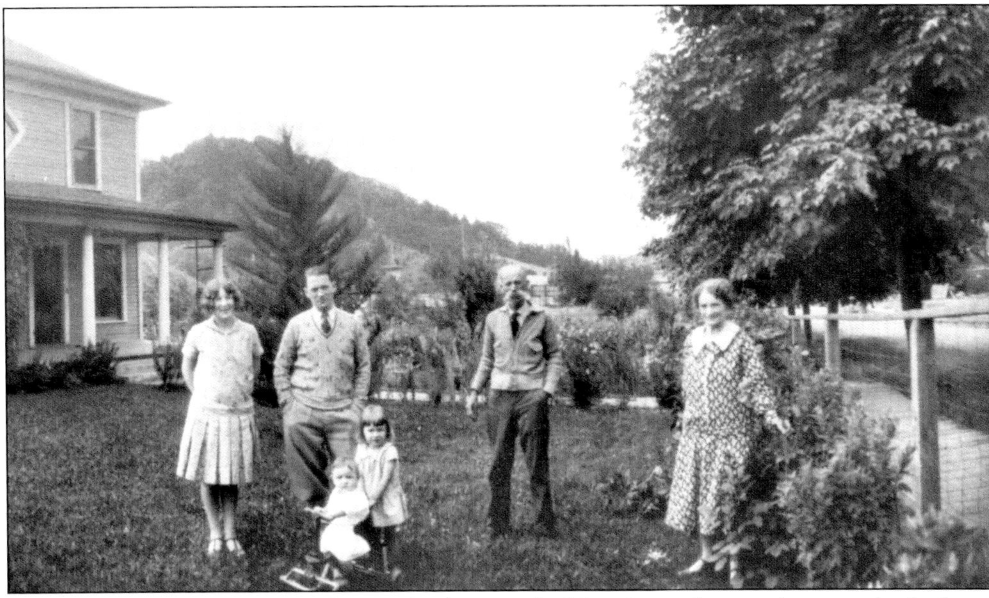

The Jackson family is seen standing in the front yard of their home (pictured opposite), which was purchased from Mayor Johns. This location continued as a boardinghouse under the proprietorship of the Jacksons. Many a bachelor logger bunked here during the timber boom of the 1940s and early 1950s. (Courtesy of Blanche Newton and Steve Hall.)

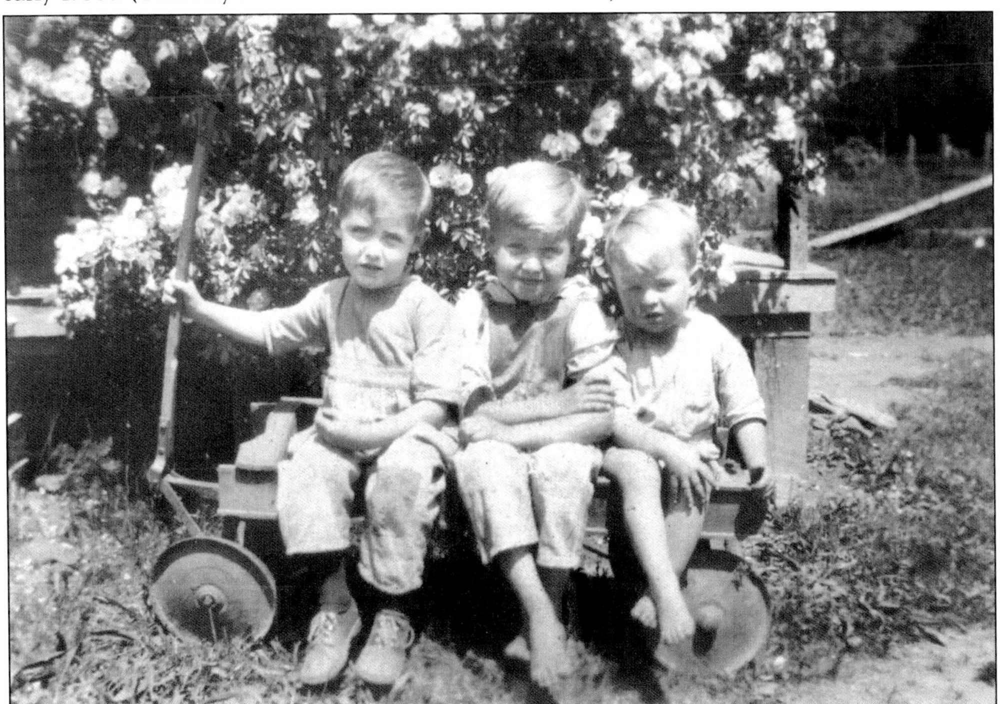

From left to right, Harry Smith and cousins Dale and Gordon Dyer pose in an old wagon in front of a blooming rose bush in this c. 1925 photograph. Harry was the son of Hank and Vida Clack Smith of Ruckles, and Dale and Gordon were the sons of Goldie and Allie Clack Dyer of Myrtle Creek. Goldie was town marshal and deputy sheriff in the 1930s. (Courtesy of Dora Smith Dyer.)

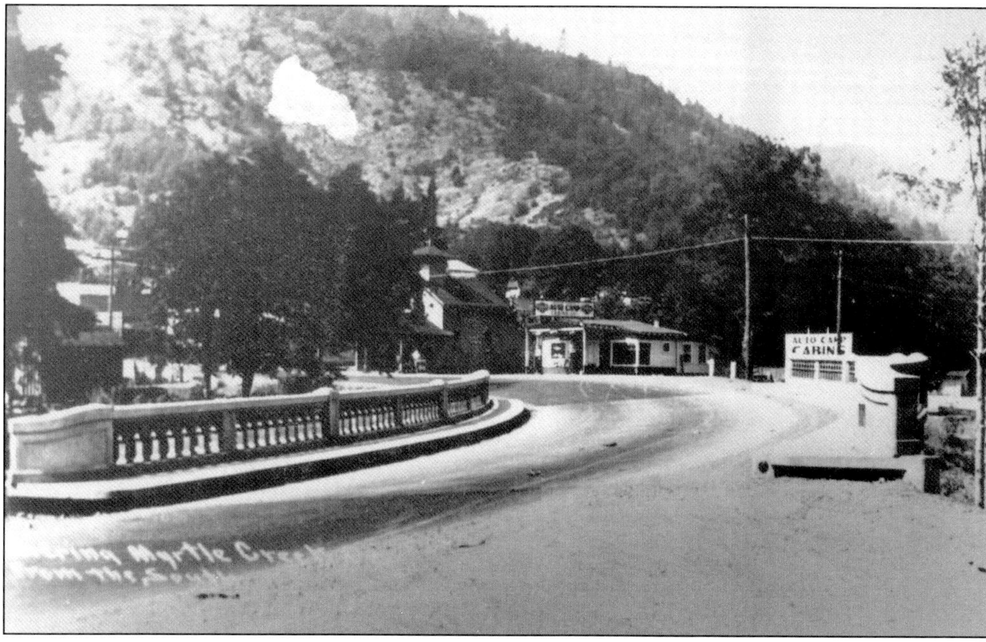

In this photograph of the bridge over Myrtle Creek, the exposed footings on the pedestrian walkway indicate that the picture was probably taken shortly after the bridge was constructed in 1930. Since 1911, the bridge had been a single-lane, steel truss structure. The removal of the landmark gristmill made way for this double-lane concrete bridge, which may have been engineered by Conde McCullough. (Courtesy of Canyonville Pioneer-Indian Museum.)

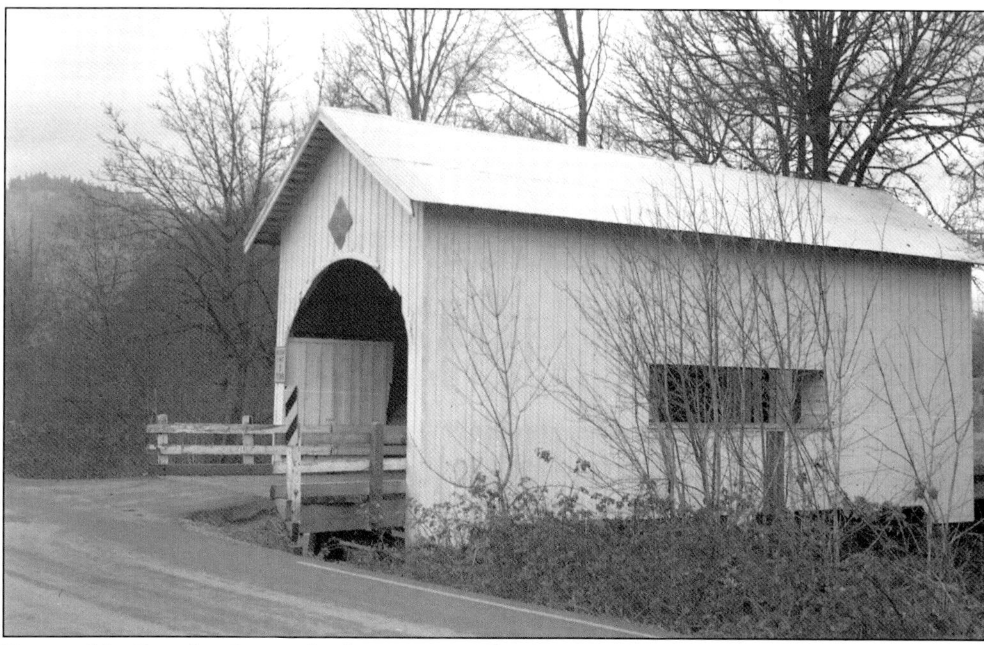

Designed by Douglas County bridge engineer Floyd Frear, who is memorialized by Frear Bridge on the North Umpqua River, the Neal Lane (or South Myrtle Creek) bridge was constructed in 1939. One of the shortest covered bridges in Oregon, this king post housed truss bridge with two approach spans is eligible for the National Register of Historic Places.

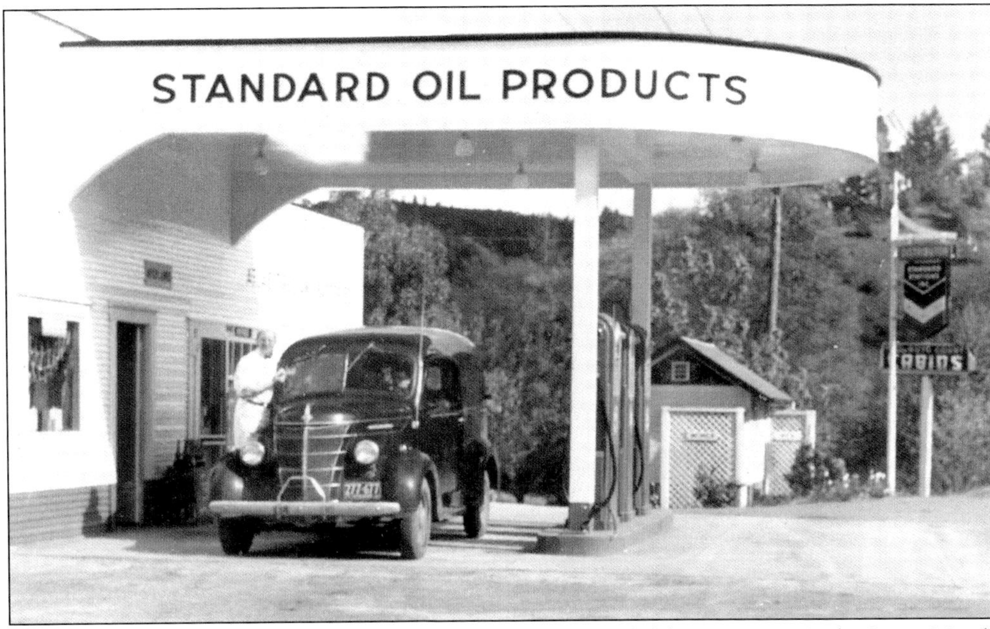

An attendant washes the windshield of a 1937 International delivery van at Myles Jones Myrtle Creek Auto Camp gas station, which was located along Main Street on a parcel currently occupied by Village Realty. The adjacent tourist accommodations were known locally as the "Jones Cabins." This service station with a mom-and-pop market was later owned by Walter Stimpson, who operated it until the late 1970s. (Courtesy of Dick Chaney.)

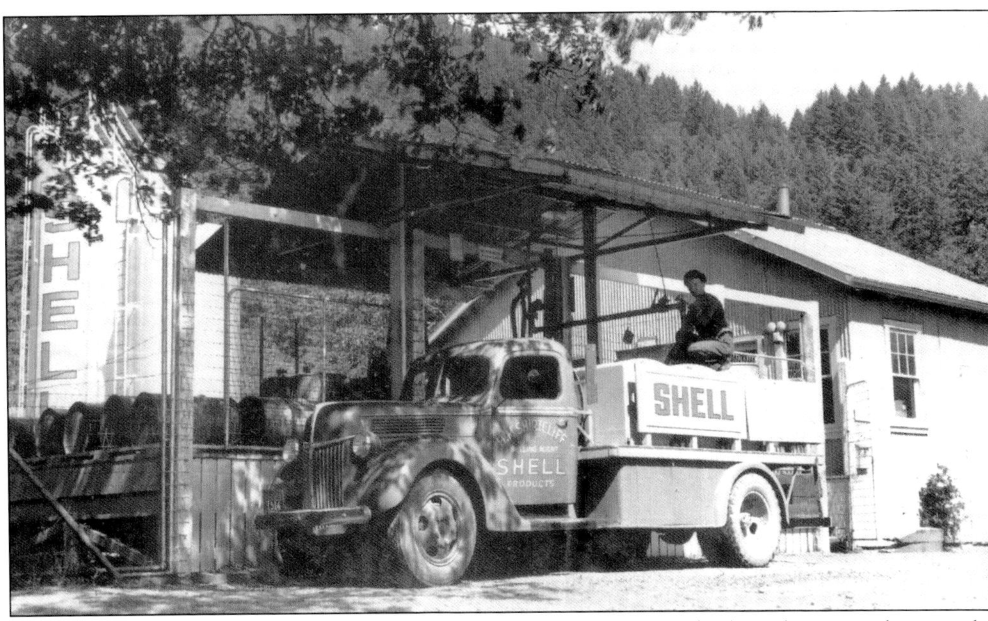

This is a 1941 photograph of the Shirtcliff Shell Oil bulk plant and oil truck. Located across the track below the railroad depot, the plant is still in business, now operated by son John Shirtcliff. (Courtesy of Dick Chaney.)

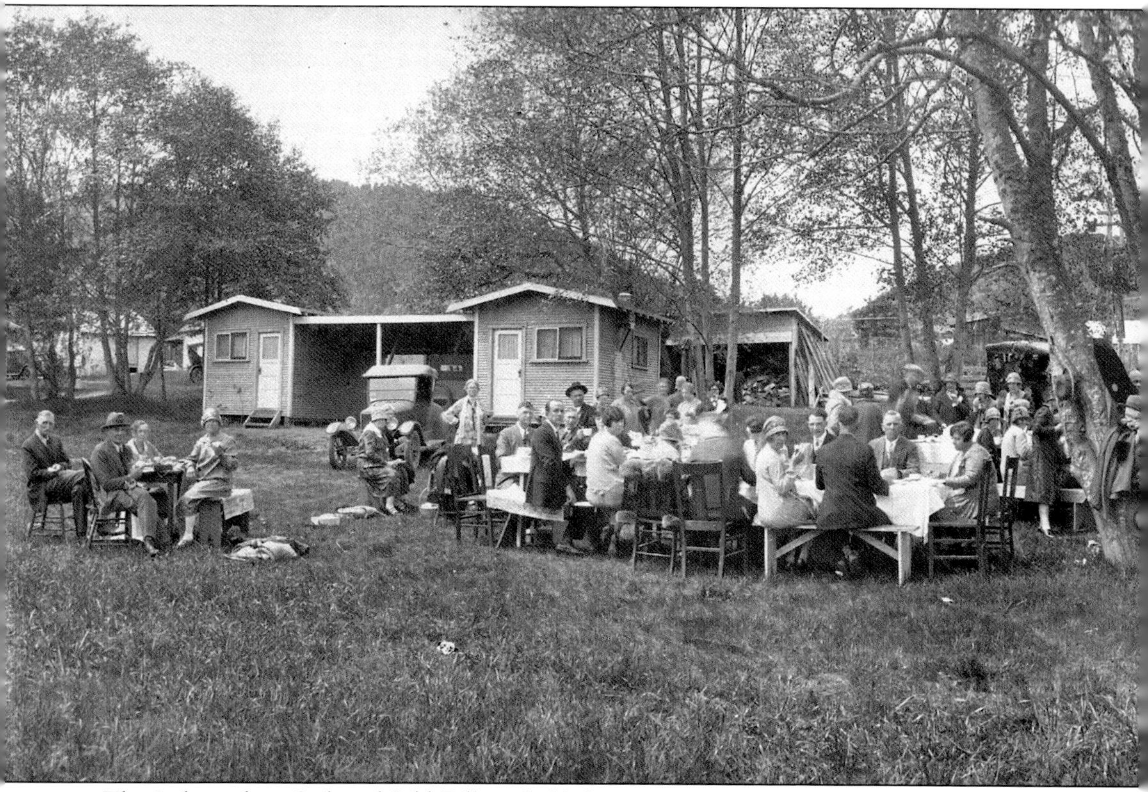

The Independent Order of Odd Fellows held their 1928 convention at the Myrtle Creek Auto Court (locally called Jones Cabins), which was located adjacent to the creek at the south end of town. This happy group is enjoying pleasant weather during their picnic under the trees. A pair of small cabins with carports between them can be seen on the far left; this was the typical "motor hotel" design of the times. On the far right, the backside of the Johns/Jackson boardinghouse

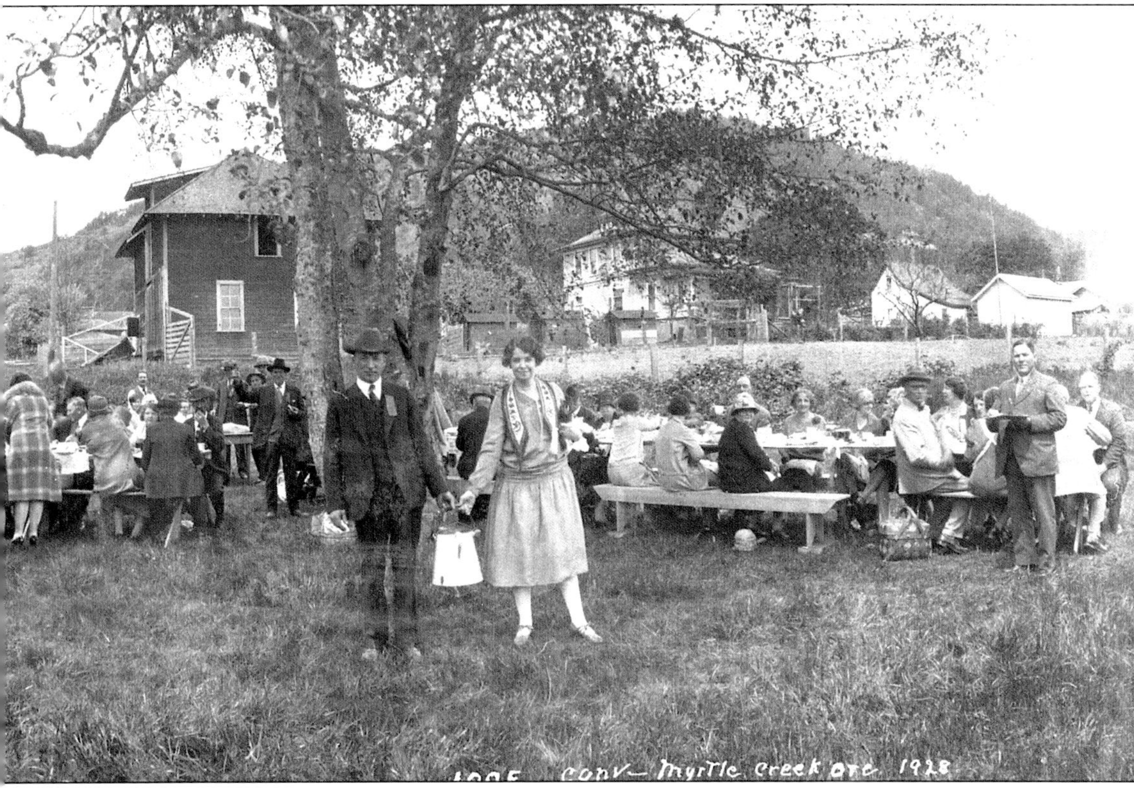

on First Avenue is visible in the distance. The couple by the tree is carrying a huge pot of coffee that appears to be "splatterware," which is an icon of the era. After a history of repeated flood damage, this property was purchased by the city with FEMA funds and turned into a city park and soccer field.

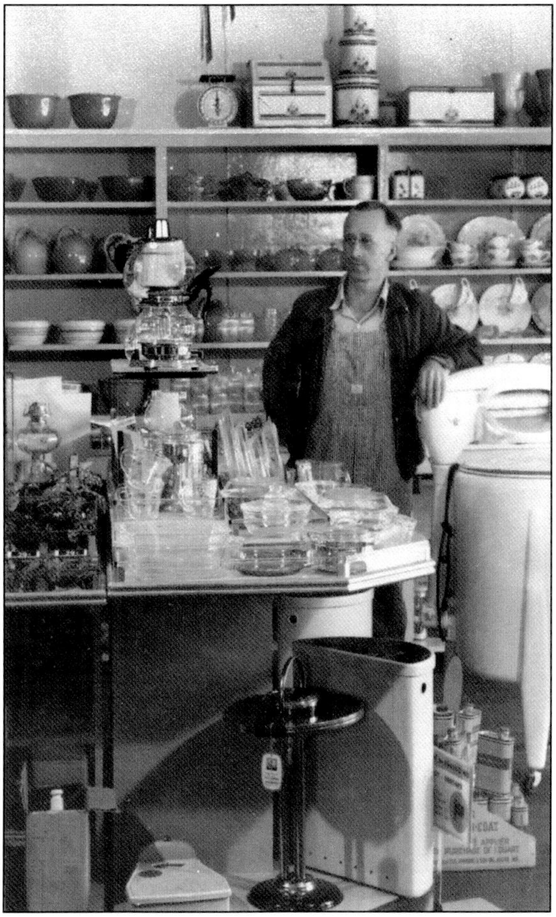

Art Verrell surveys the 1941 merchandise available in his business, Douglas County Flour Mill #2. To call him, one rang the operator and asked for number 235. You could then ask about the Fiesta Ware on the shelf, the nice smoking stand, the white enameled steel kitchen trash can, or you could delight him by asking about the fancy new wringer washer he is leaning on. (Courtesy of Dick Chaney.)

Located where the Penny Pincher now stands, a Mr. Kerby and his family lived beside their store. One night, his eight-year-old son was pumping gas when the lantern tipped over, igniting the gas fumes. A passenger in the car was badly burned, but the burns suffered by the son cost him an ear and a badly scarred face. The following weekend, Easter Sunday 1929, the Division Street school burned.

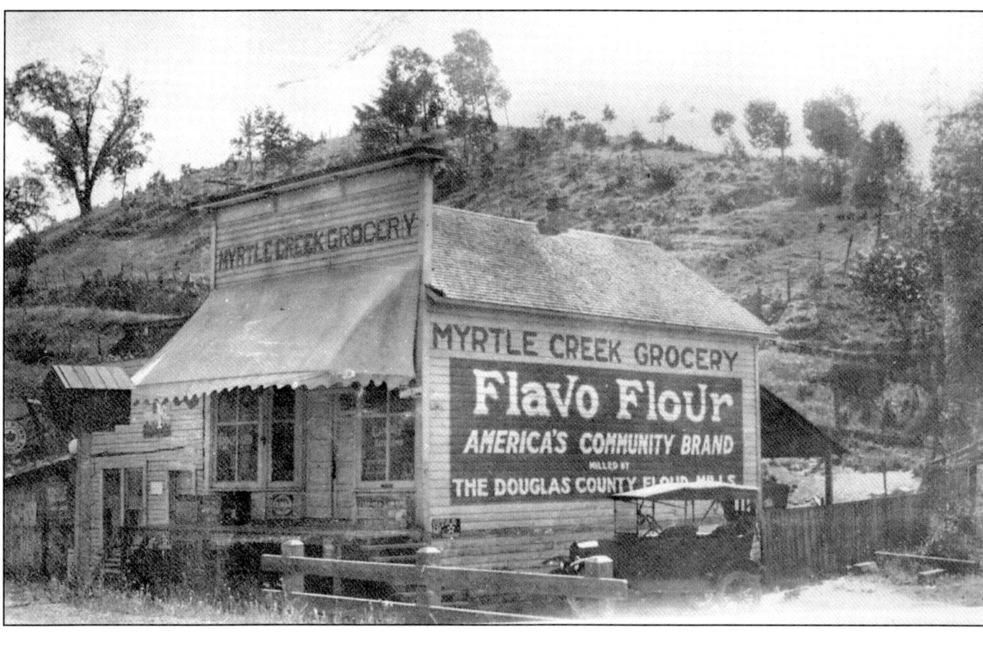

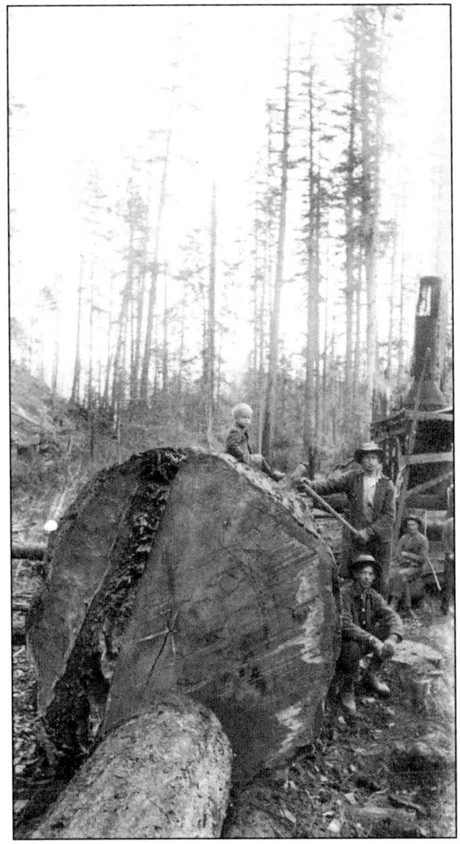

With his axe in hand and his young son on the log, Jesse Smith and his logging crew take a break in front of the steam donkey. After yarding this prime, old growth Douglas fir log to the landing, they proudly posed for the camera. (Courtesy of Dora Smith Dyer.)

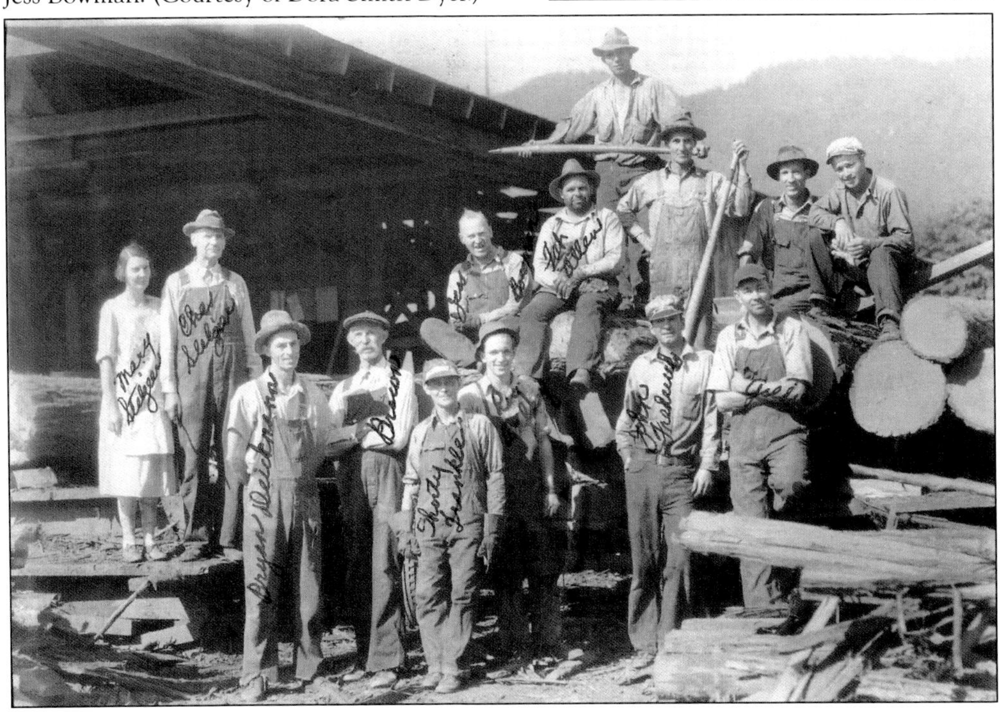

It started in 1923 as a box mill, but by the time this 1937 photograph was taken, the Stelzier Mill had grown into a full-fledged sawmill. Shown here at the Johnson Street site about two blocks down from the current post office are Mary and Charles Stelzier posing with crew members of the sawmill. Among them are Bryan Dieckmann, Shorty Trankle, Roy Anlauf, Goldie Dyer, and Jess Bowman. (Courtesy of Dora Smith Dyer.)

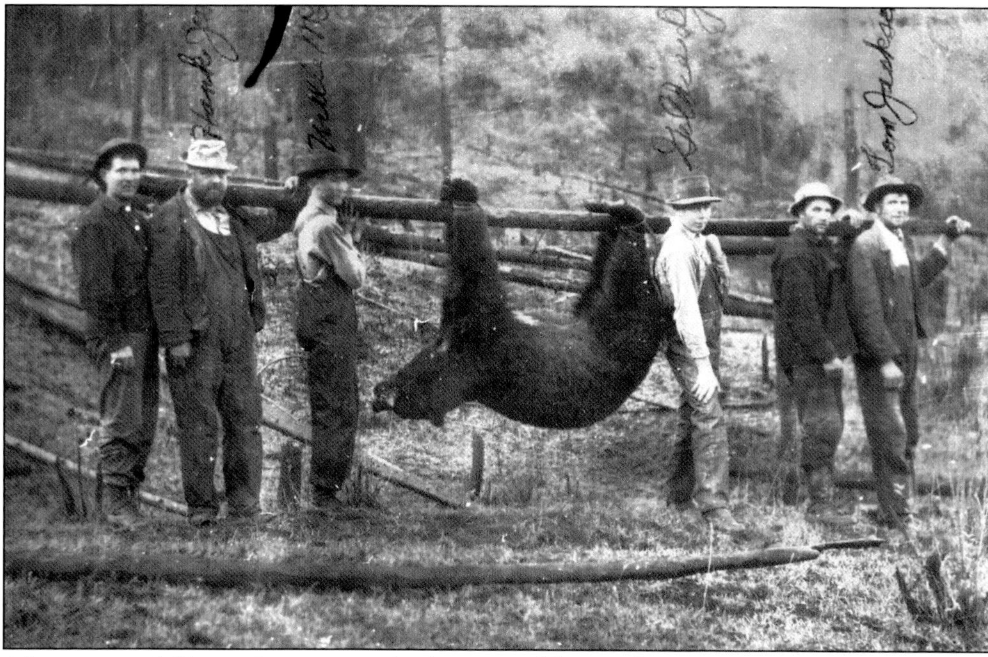

After a successful hunt, these men pack a bear out of the woods. During the Depression, hunting wild game was practiced by most rural citizens and provided a valuable source of protein to the diet. The happy hunters here are, from left to right, unidentified, Hank Jones, Willie McClain, Goldie Dyer, unidentified, and Tom Jackson. (Courtesy of Dora Smith Dyer.)

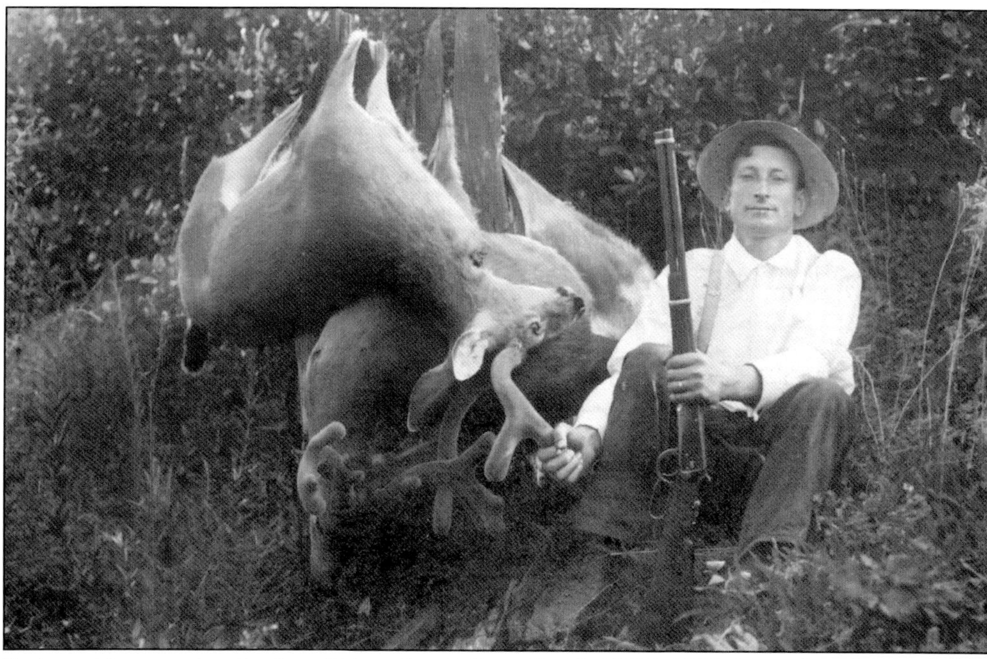

Posing with two prime bucks still in velvet is Jesse Smith (1884–1972). His rifle is a Winchester model 1894 30-30 saddle-ring carbine with a tang peep sight. Selling for about $20, this was a very popular rifle and often provided the only meat many local, Depression-era families had to eat. (Courtesy of Dora Smith Dyer.)

Seven

Timber Boom

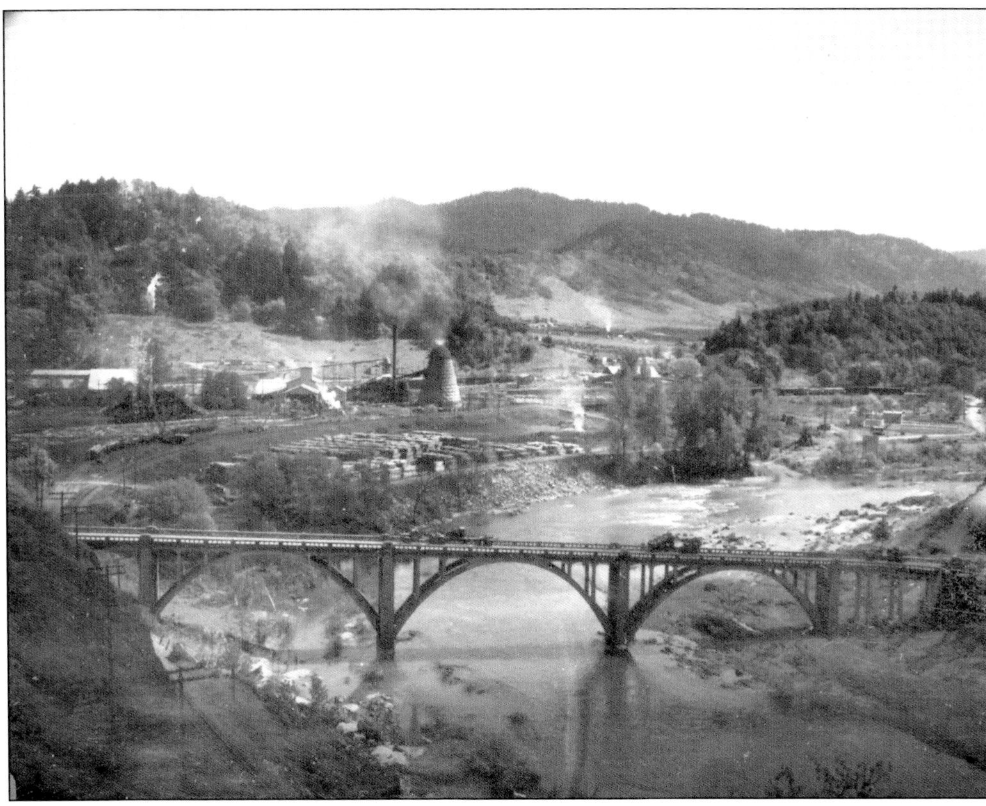

With the coming of the post–World War II timber boom, both of Myrtle Creek's entrances into town—the railroad and Pacific Highway 99—were bustling with activity. The bridge was lined with trucks hauling logs in, and the tracks were lined with railcars hauling lumber out. (Courtesy of City of Myrtle Creek.)

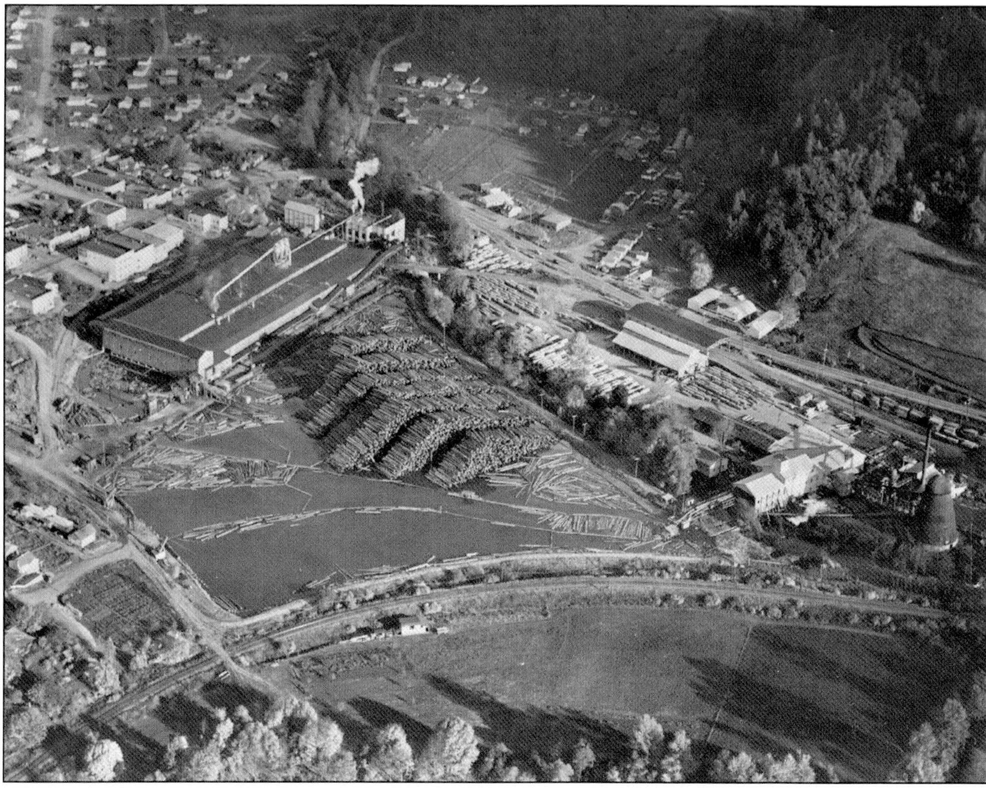

This c. 1950 photograph shows the Umpqua Plywood Corporation with full, cold decks of logs, log rafts boomed up in the pond, and a railroad siding that crossed the creek and ran along the mill. Landmarks not yet built are the Dairy Queen and Winetrout Ford, now Don Jones Automotive. (Courtesy of City of Myrtle Creek.)

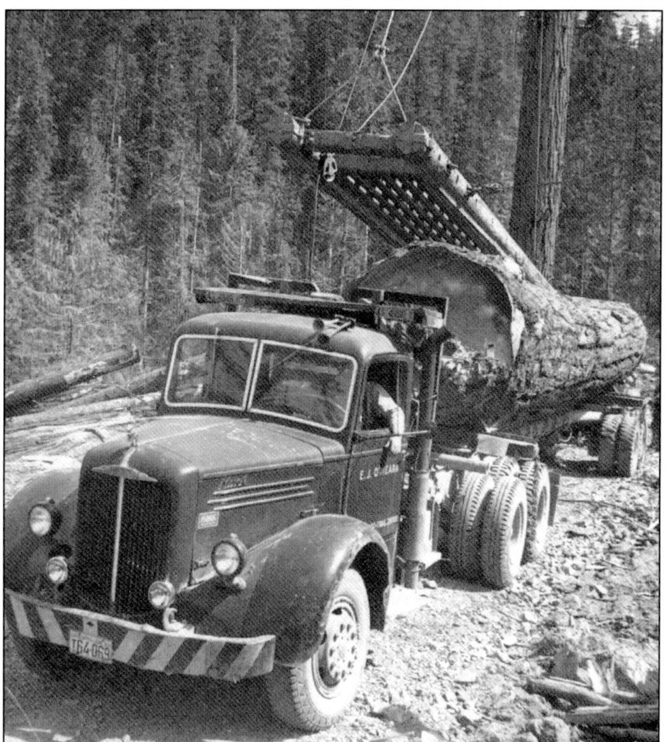

Squatting under the weight of a one-log load placed on the trailer by the heel-boom log loader at the rear, E.J. O'Meara and his GMC Mack truck will soon be headed down the mountain to the mill. (Courtesy of Dora Dyer.)

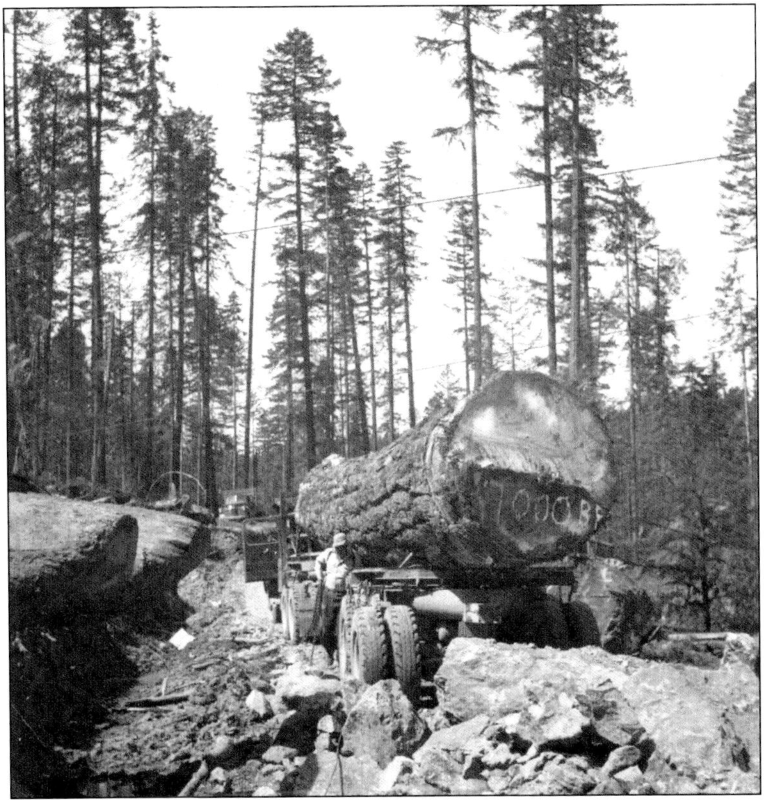

With a chain and binder slung over his arm, a driver prepares to chain the log to the trailer and cinch it down with a binder and cheater pipe. Two or three more wraps and he will be ready for the arduous trip down the mountain. On the way down the steep grades, drivers crossed their fingers and prayed that their brakes did not fail. (Courtesy of Dora Dyer.)

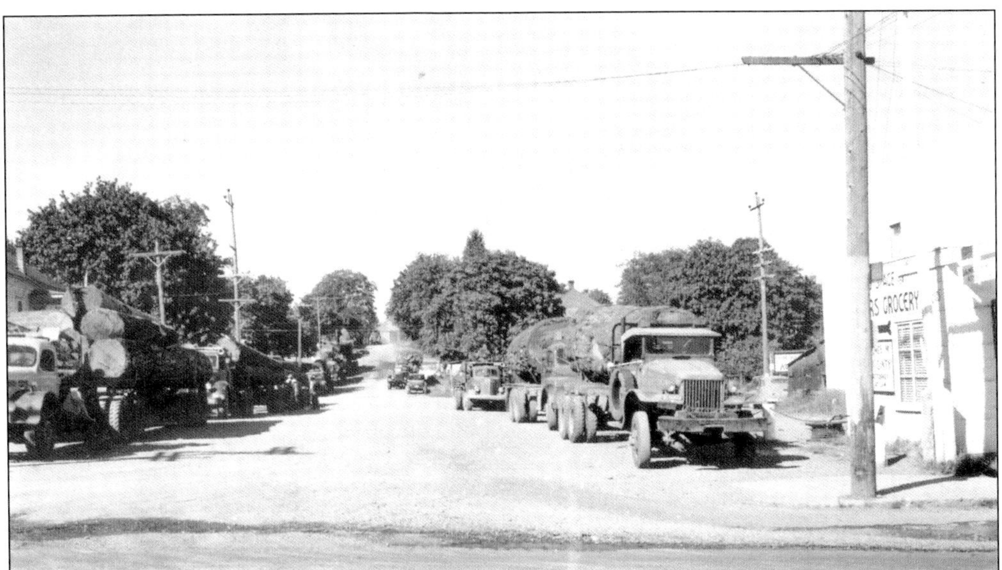

A queue of post–World War II log trucks, several with one-log loads, waits on Third Avenue for their turn at the log dump. On the right is a military surplus "deuce-and-a-half" with a heavy-duty canvas cab roof for the ultimate in driver comfort and protection. (Courtesy of City of Myrtle Creek.)

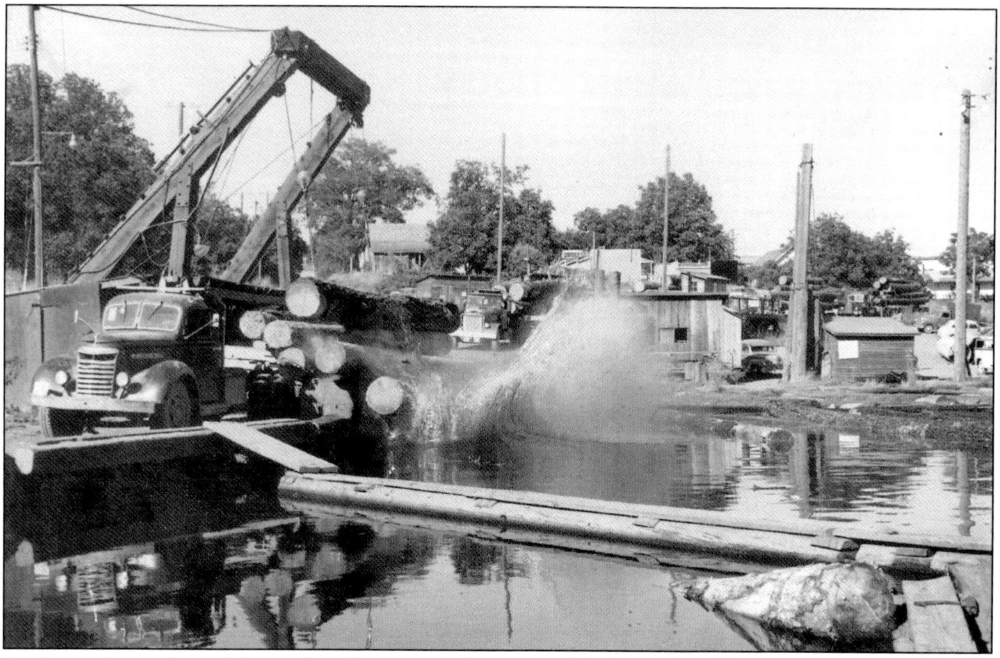

At the mill pond, an A-frame log dump rigged with a cable drum winch empties the logs into the pond and then hoists the trailer up on the back of the truck. With the trailer chained down, the driver is ready for the long trip back up to the landing for another load of logs. Depending on the distance from the landing, a driver usually made from two to four "turns" a day.

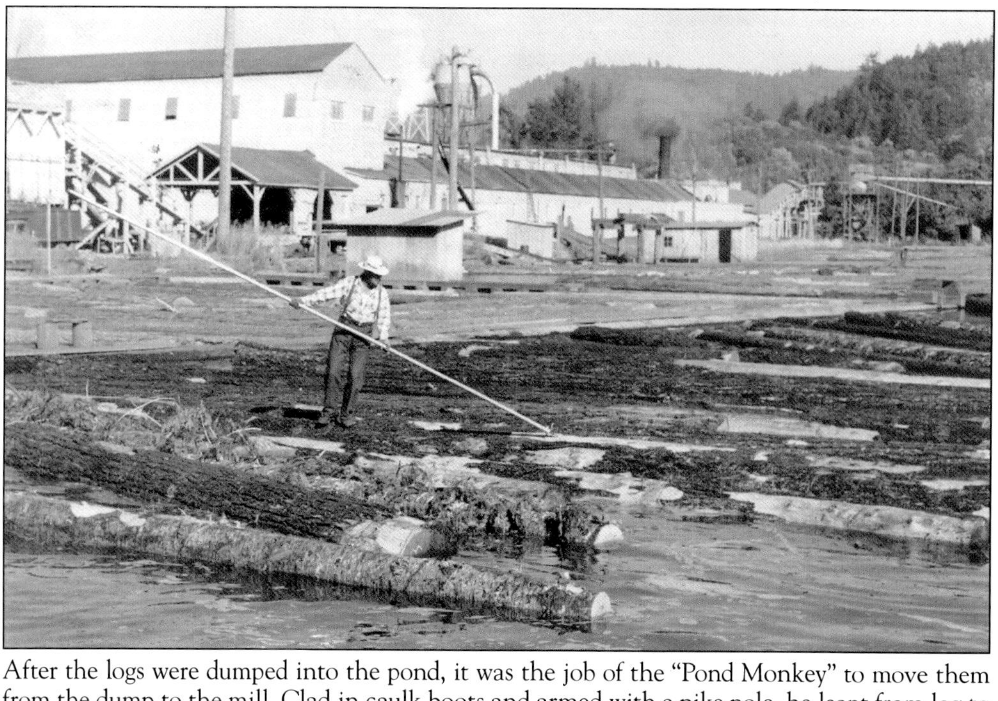

After the logs were dumped into the pond, it was the job of the "Pond Monkey" to move them from the dump to the mill. Clad in caulk boots and armed with a pike pole, he leapt from log to log and maneuvered the logs across the pond towards the mill. It was a job requiring strength, coordination, agility, and a "devil-may-care" attitude about taking a dunking.

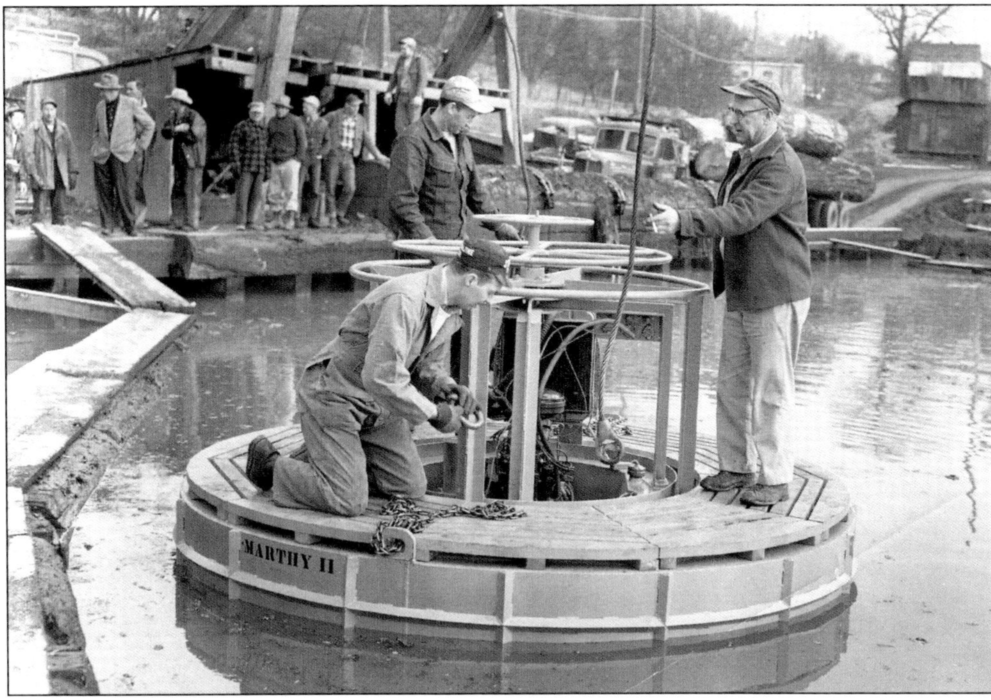

When the pond monkey was unable to move the logs, a pond boat was available to help him out. Normally, pond boats were a small, heavily constructed scow with an outboard motor center mounted in a well. This 1951 photograph shows a modern, high-tech version being assembled on the Firmco pond in Myrtle Creek (present location of Millsite Park.).

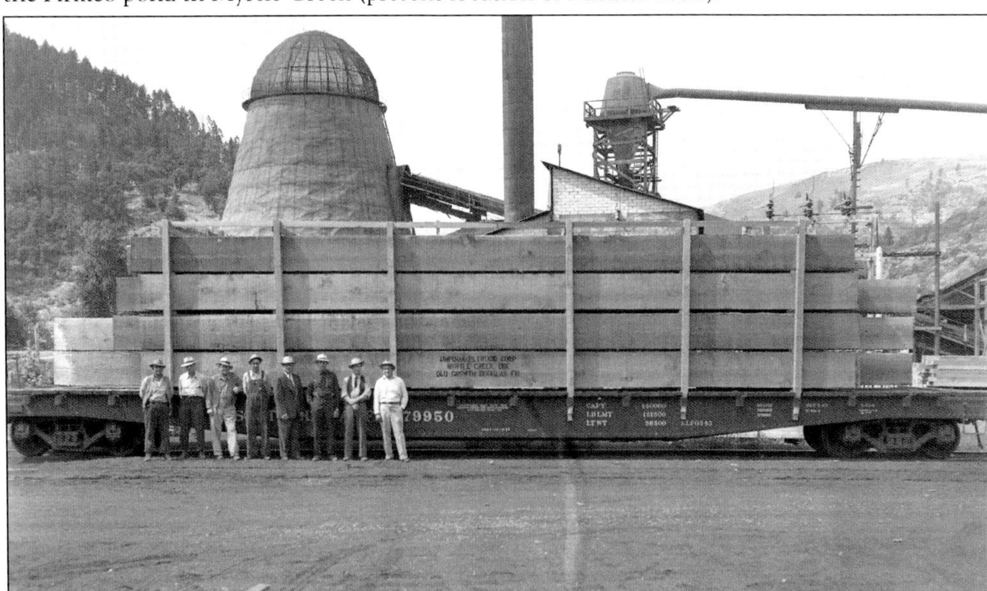

Stamped "Umpqua Plywood Corp. Myrtle Creek, Ore, Old Growth Douglas Fir," these huge, one-piece beams are loaded and ready for shipment. The sawmill was able to cut 90-foot logs and on occasion shipped beams that required two railroad cars fitted with a fifth wheel to transport. Obviously custom-cut, it would be interesting to know where these beams were headed.

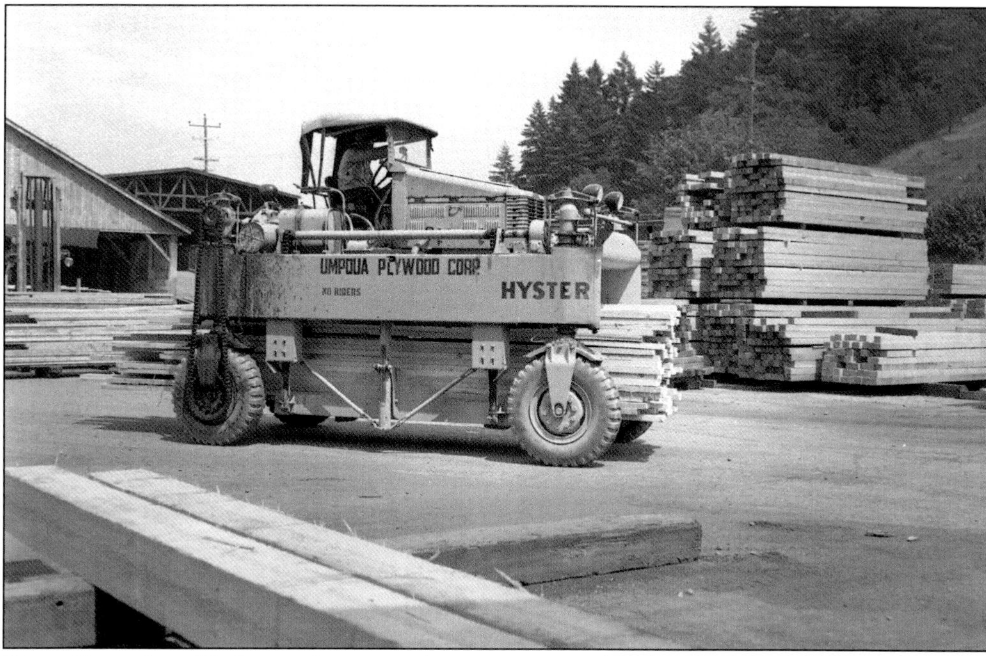

These Hyster "straddle buggies" were used to transport stacks of lumber around the mill yard. Up until the late 1970s, every 17 minutes the Green Valley Mill on North Myrtle Road sent one of these through town to the rail siding on South Main Street.

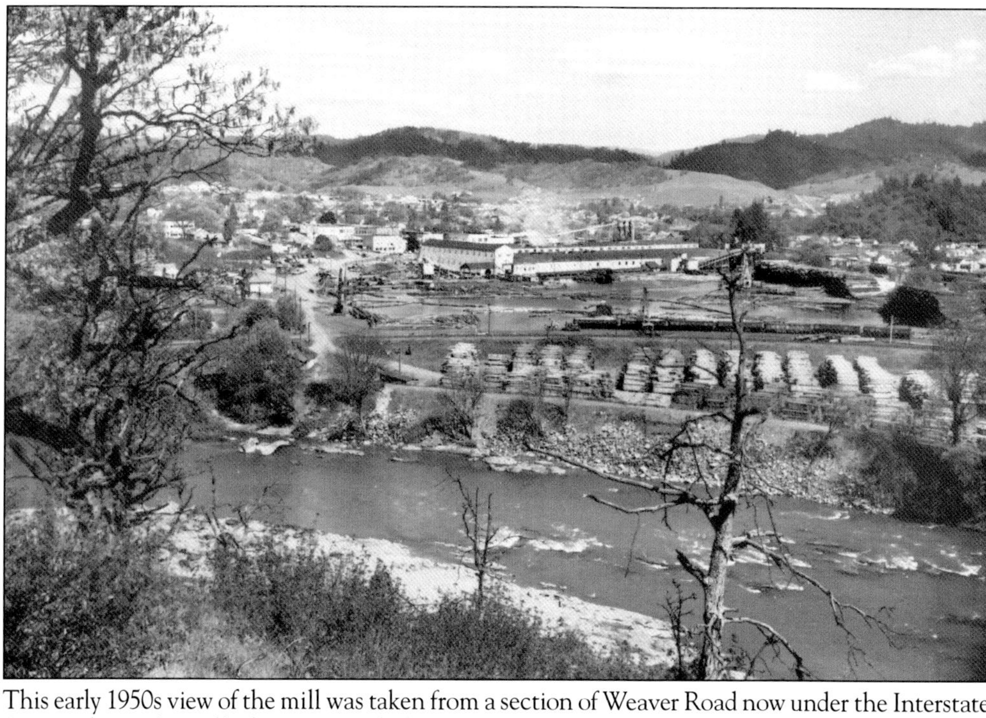

This early 1950s view of the mill was taken from a section of Weaver Road now under the Interstate 5 exit. This is the mill when it was a hub of activity and in full production. There are cold decks of logs and stacks of lumber along the river, a rail siding along the log pond, and Myrtle Creek was at the height of its timber boom. (Courtesy of City of Myrtle Creek.)

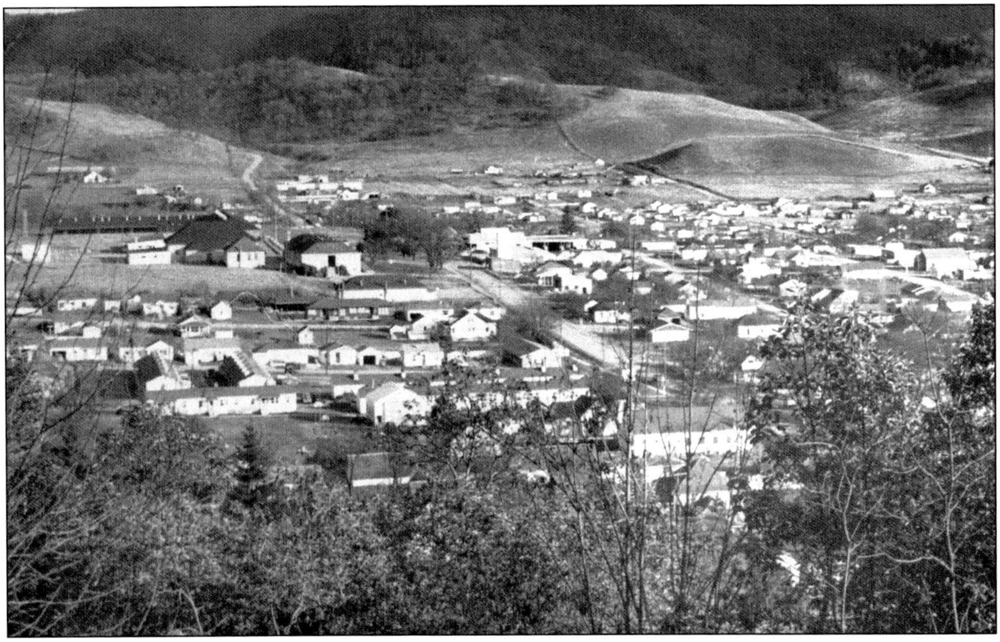

The population boom following World War II created a housing shortage, forcing many local families to live in tents. A Federal Victory Housing Project, known as Myrtle Manor, was built along Bataan Avenue. It is the lower left group of buildings in this photograph. It was torn down in the early 1950s, and the City of Myrtle Creek purchased the land. A swimming pool, library, and skateboard park now occupy the site.

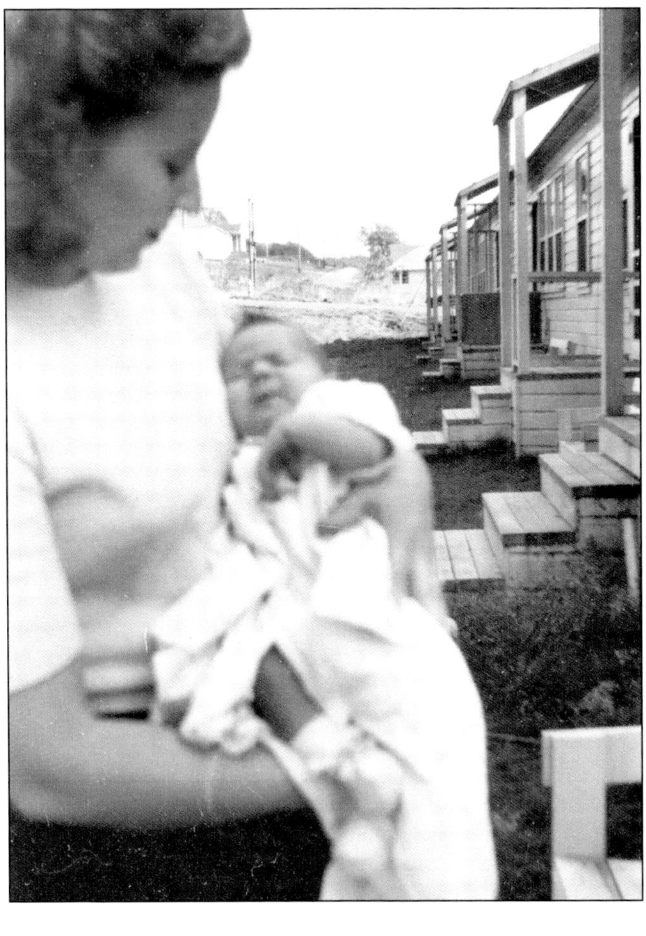

In 1948, Betty Briggs Hanna shows off her daughter Bonnie in front of the Victory Housing Project on Bataan Avenue. These compact units were built on government land to house the many homeless families after World War II. A high school—built on Rice Street in 1946 and turned into a junior high in 1965—is visible in the background. (Courtesy of Becky Hanna Norton.)

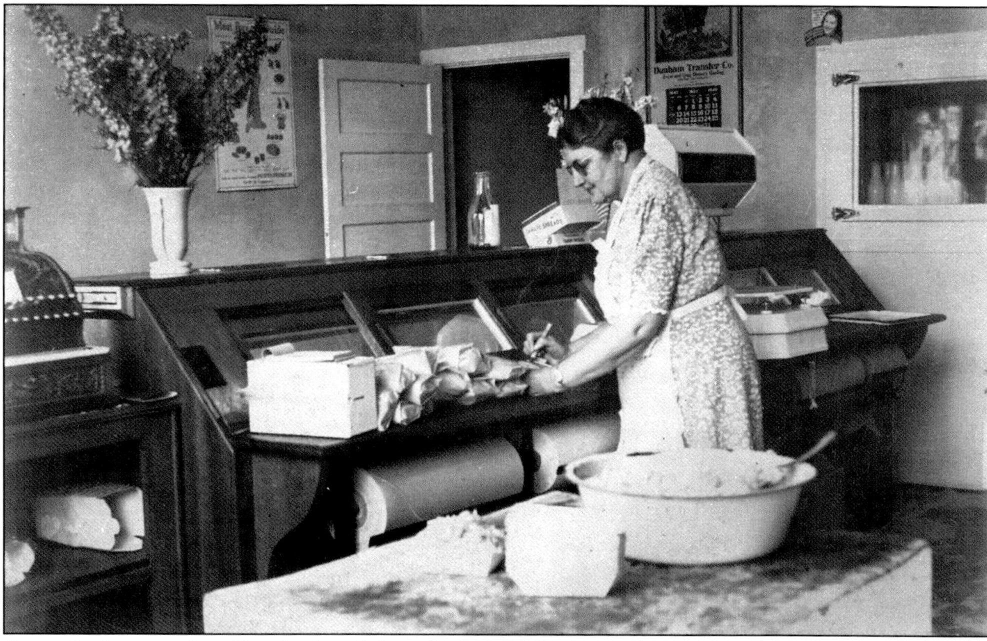

Emma Ledgerwood and her husband, Joseph, started their meat business in 1922 at the Ledgerwood Ranch on Boomer Hill, where they raised and slaughtered cattle. In 1926, they purchased Sank Buell's butcher shop on Second Ave. In 1940, the City Meat Market moved to Main Street, where the scale was new, but they still used the ornate cash register that was in Sank's shop.

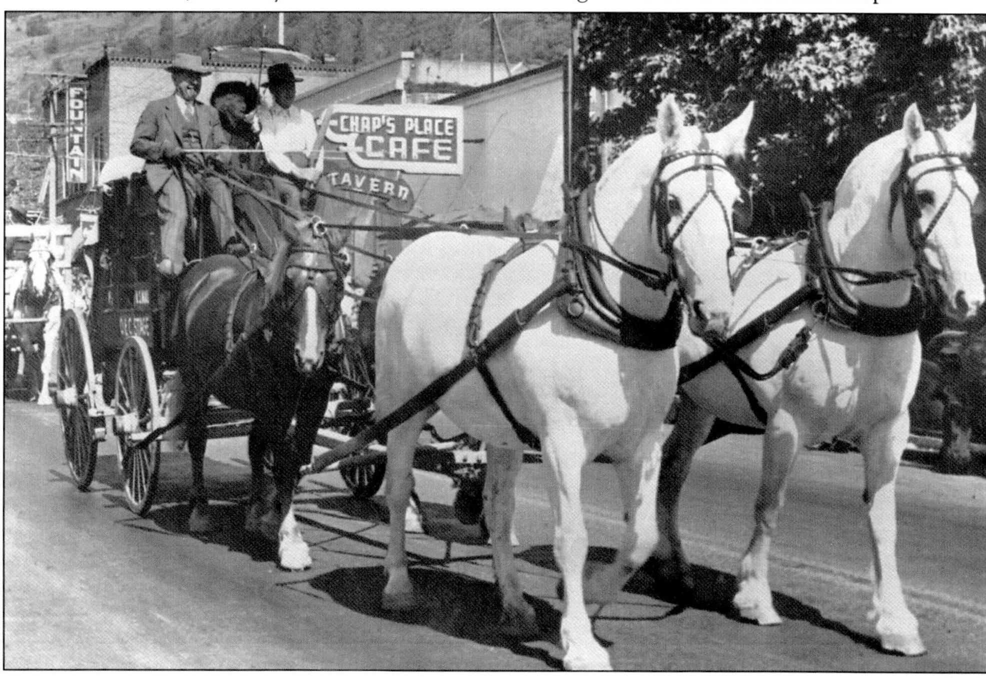

In this c. 1950 photograph, George Burnett drives his father Tom's stagecoach in a parade down Main Street in front of Chap's Place. An art now nearly lost, driving a multi-horse team is more difficult than it seems at first glance. It requires practice and horses trained to react properly according to their position in the team.

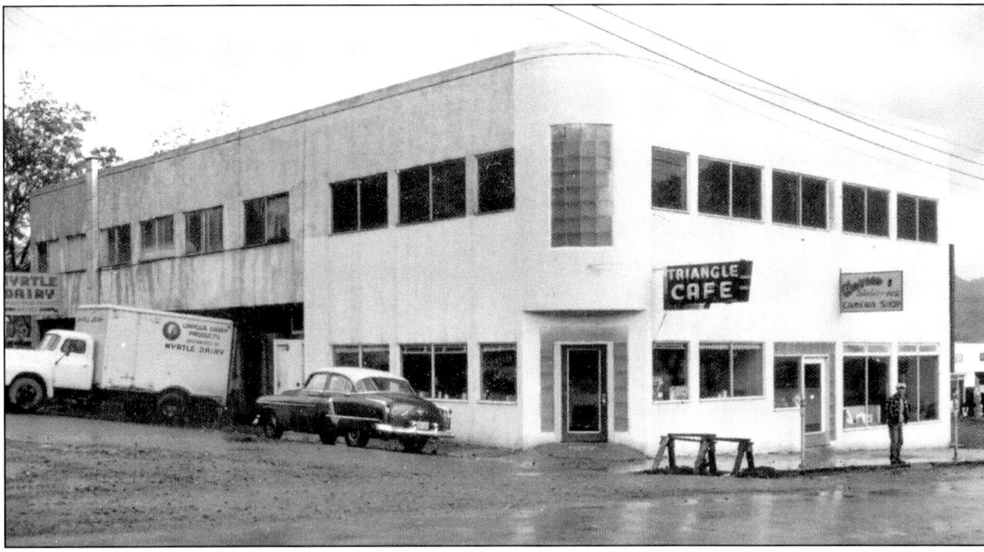

Art Moderne, a style of Art Deco architecture, is prominent in many buildings constructed in the Myrtle Creek business district during the second quarter of the 20th century. The hotel at Main Street and Fourth Avenue is a fine example of the curved walls, glass-block windows, and minimal embellishment of Art Moderne. The Triangle Cafe, Warren's Photo Studio, and Myrtle Dairy occupied the main level in the 1950s, when this photograph was taken.

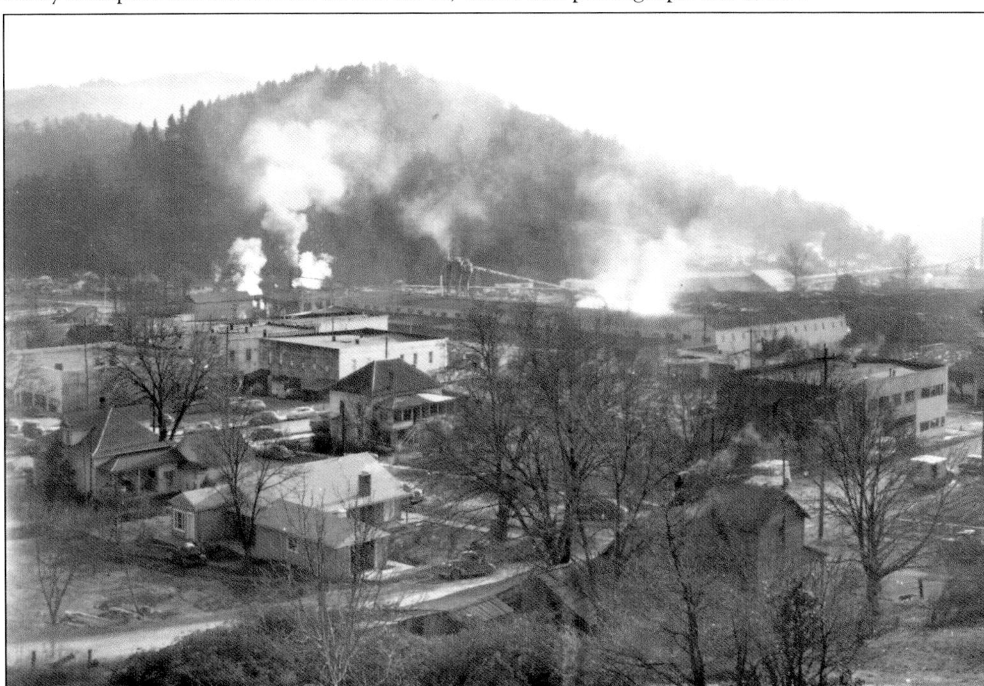

This c. 1950 view to the south over Myrtle Creek is imbued with the ambiance of a typical southern Oregon postwar timber boomtown. With stacks emitting fumes and steam and wigwam burners belching smoke and sparks, these towns smelled heavily of wood smoke and log pond aromas. During weather inversions, the smoke and fog settled into the valley and made the air "as thick as pea soup." (Courtesy of City of Myrtle Creek.)

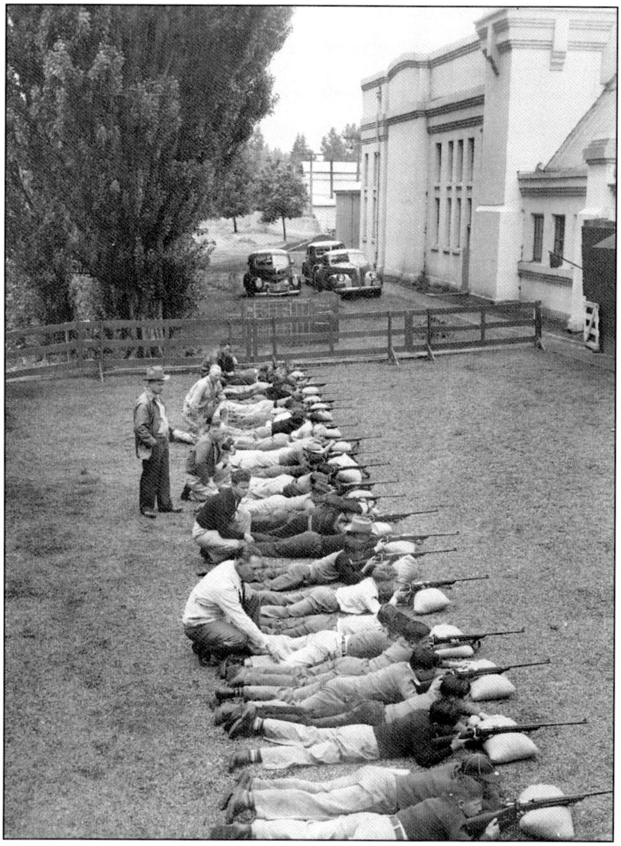

In this June 1941 photograph, the Myrtle Creek High School rifle team practices behind the Division Street School under the tutelage of principal and coach Tom Ireland. The high school also boasted a girls' rifle team with as many members as the boys' team. Six months later, Pearl Harbor was bombed, and it can be assumed that some of these students soon had to put their rifle skills to use.

Umpqua Research now occupies this building at the top of Division Street across from the grade school. The Goal Post opened around 1950, and until the high school was moved to Tri City in 1965, it was a popular lunch hour destination for students from the nearby high school. (Courtesy of City of Myrtle Creek.)

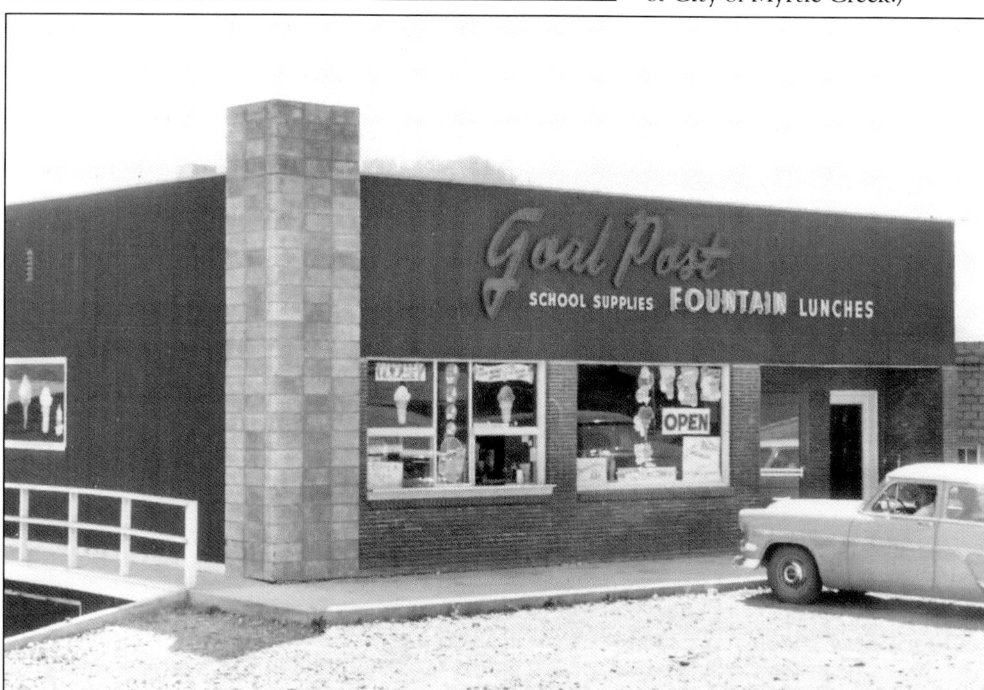

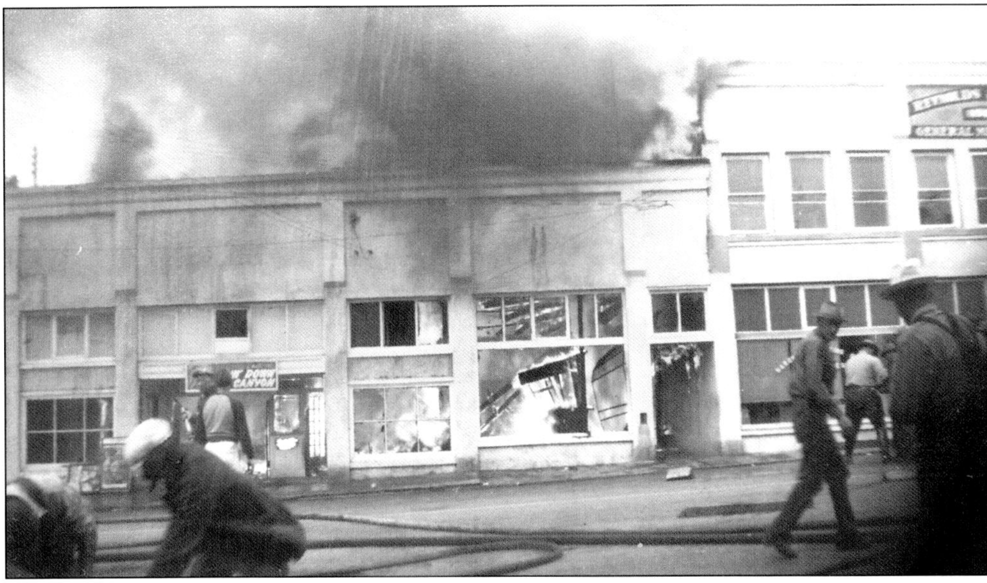

This 1943 fire on the west side of Main Street destroyed nearly a block of businesses. Fully engulfed in flames at the center are the remains of the post office. Down the street, the City Meat Market and Bakery and an unknown business next to the post office met a similar fate. (Courtesy of Betty Hanna.)

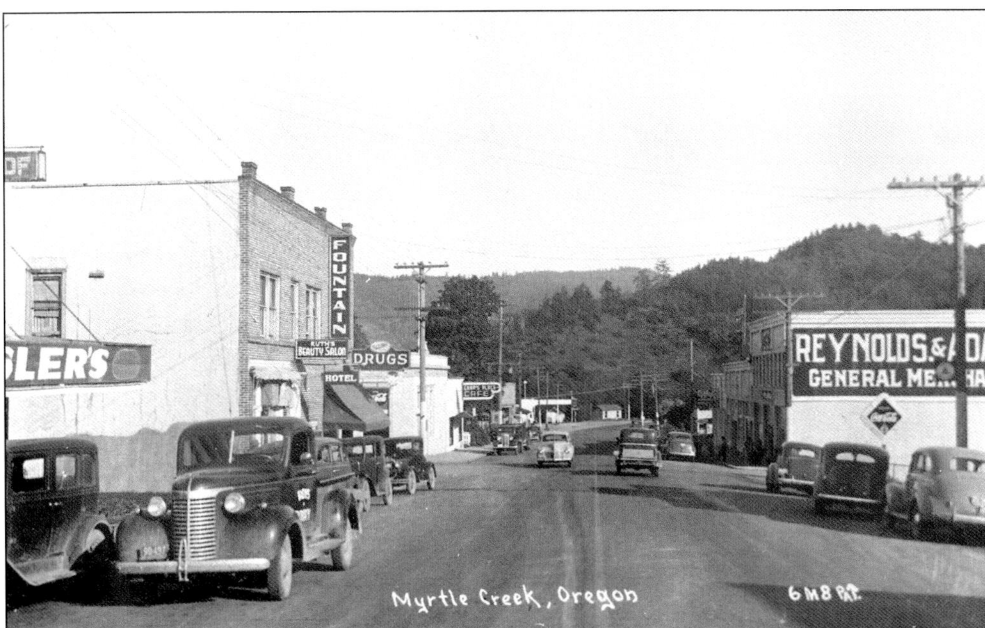

A 1940 Chevrolet pickup is double parked in front of Kusler's Market in this view looking down Main Street. Across from the Terminal fountain, café, and Greyhound bus depot is the Reynolds and Adams store in the Rice Brothers and Adams building. Ted Rice has moved Myrtle Drugs to its new location across Second Avenue from the Terminal. Passing by the drugstore is a 1937 Hudson Terraplane.

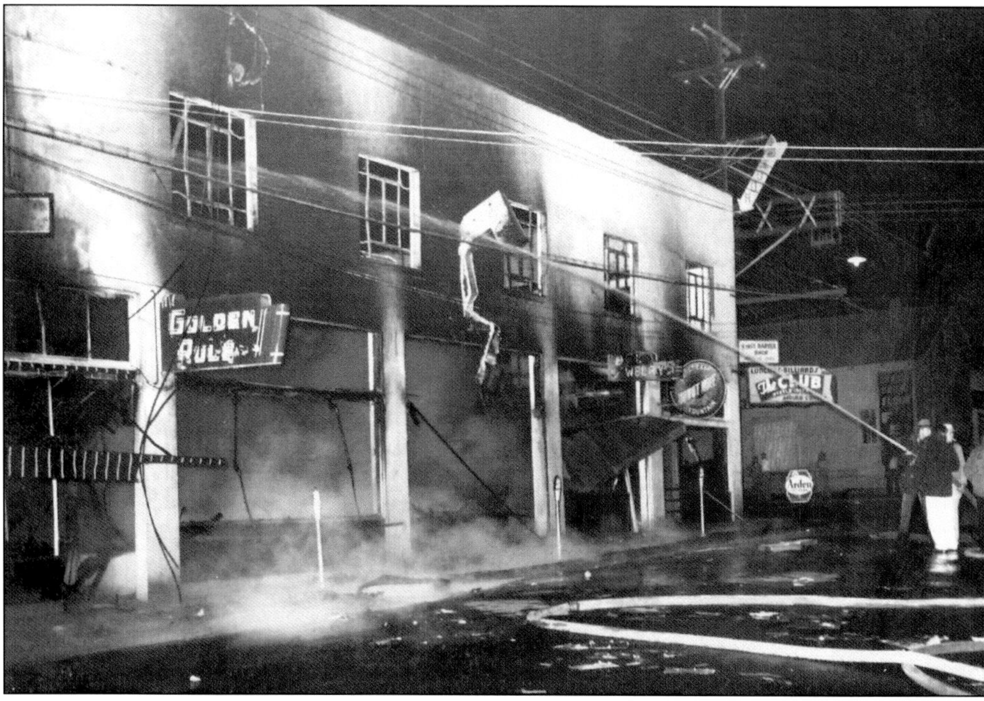

On September 29, 1950, a fire broke out at 10 p.m. and gutted the buildings along Second Avenue between Main and Oak Streets, forcing the occupants of Palmer's boardinghouse to leap from the second-floor windows. One man perished, and 15 businesses were lost. Impacted stores and offices included Kelly's Furniture, Gosline's Jewelry, Umpqua Florists, Golden Rule, Dr. Parson's Chiropractic office, the Club, the Do-Nut Hole, a barbershop, an auto parts store, two dentists, and several more. Most businesses never reopened; however, the Golden Rule relocated across the street and became one of the most popular clothing stores in town. The photograph above shows firefighters on Second Avenue, and below is an Oak Street view.

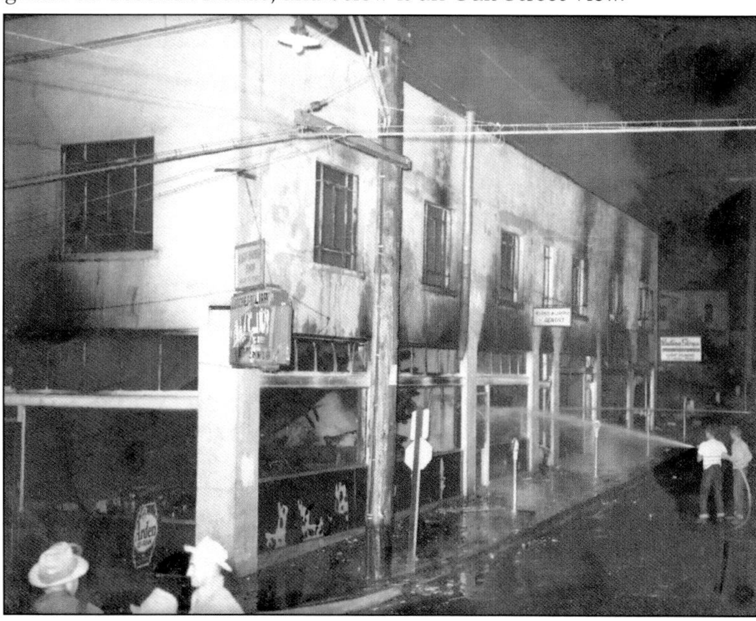

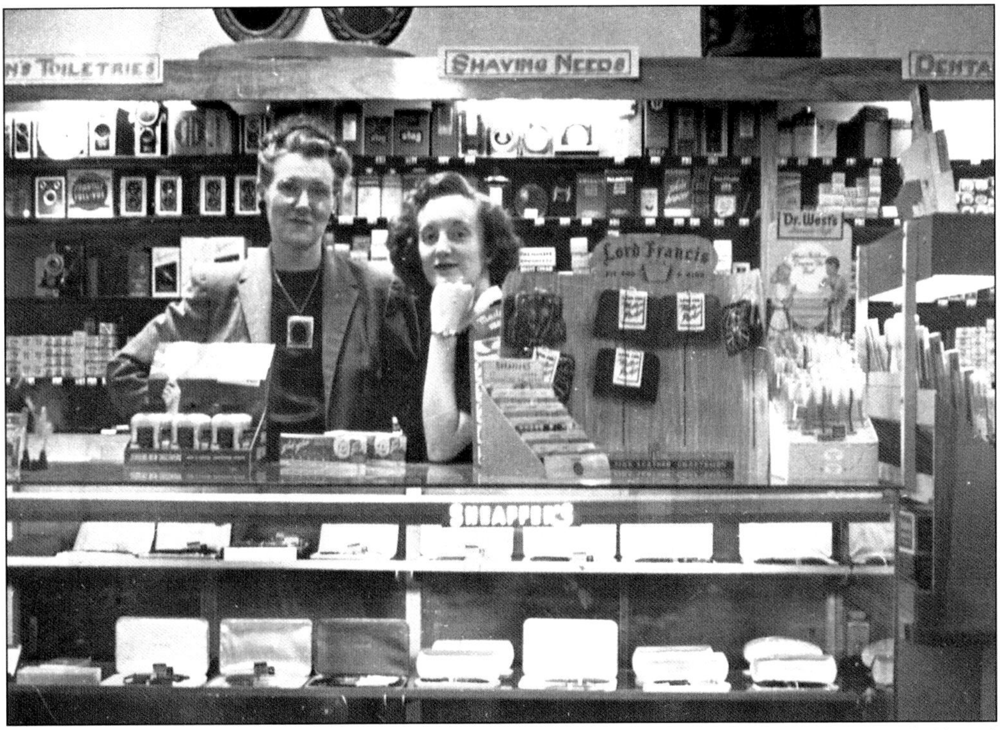

In this 1948 photograph of the Rice Brothers Myrtle Creek Drugstore, clerks Milly Fogus (left) and Audrey Krause pose over a glass display case featuring Sheaffer's fountain pens. Beside Audrey, Lord Francis "Fit for a King" leather handbags are arrayed, and on the rear wall are toiletries, shaving supplies, and dental care products.

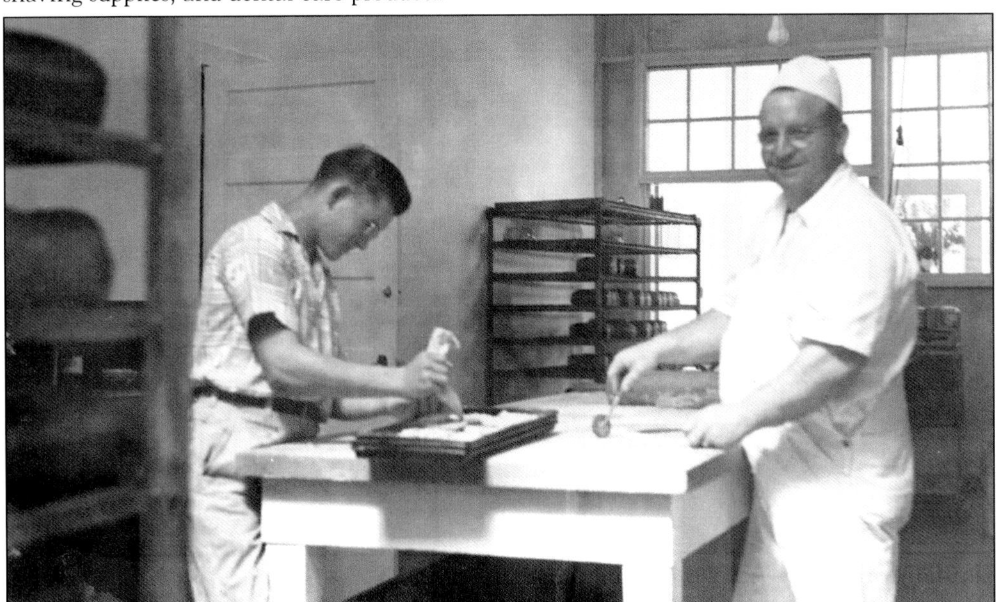

Mr. "Mac" McDougal cuts dough as his son Harry ices a cake in the Myrtle Creek Bakery in this 1941 photograph. Before supermarkets, food shopping involved a stop at a grocery store, a butcher shop, and a bakery to replenish the pantry and icebox. (Courtesy of Dick Chaney.)

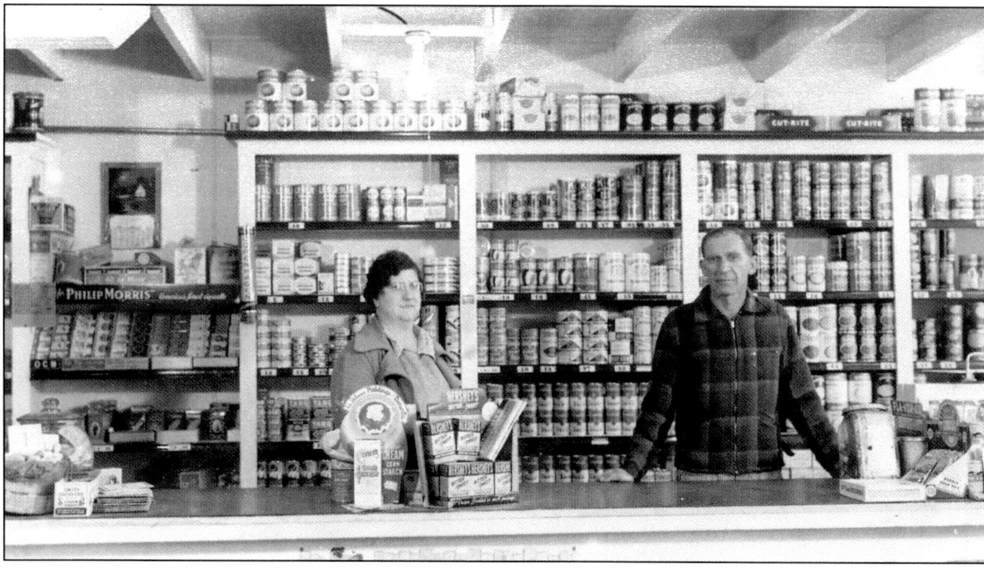

Mr. and Mrs. Easton are pictured in their grocery store on Second Avenue in 1941. Closing this store, they opened a smaller store on the lower corner of Spruce and Division Streets. Called the Snack Shack and with a stock of lunch, soft drinks, and snack items, it was a popular destination for the school kids across the street. Later, Mildred Walker and Lou Pringle took over management of the Snack Shack. (Courtesy of Dick Chaney.)

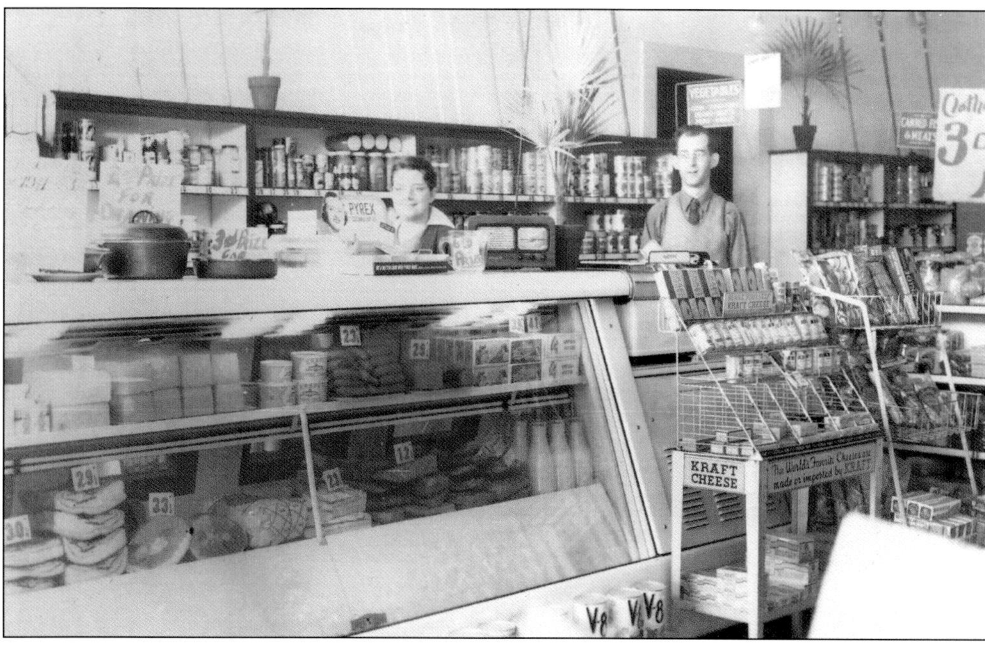

LaVonna Harris Kusler and an employee are pictured in Kusler's Super Mart on Main Street. Around 1930, Marion Kusler purchased the interest of his father, L.M. Kusler, in the Kusler Cash Store and operated a small grocery store on the ground floor of the IOOF-owned Moore building. Expanding into adjacent space, Marion and LaVona remodeled the store into Myrtle Creek's first supermarket. (Courtesy of LaVona and Allan Kusler.)

Eight

Changing Times

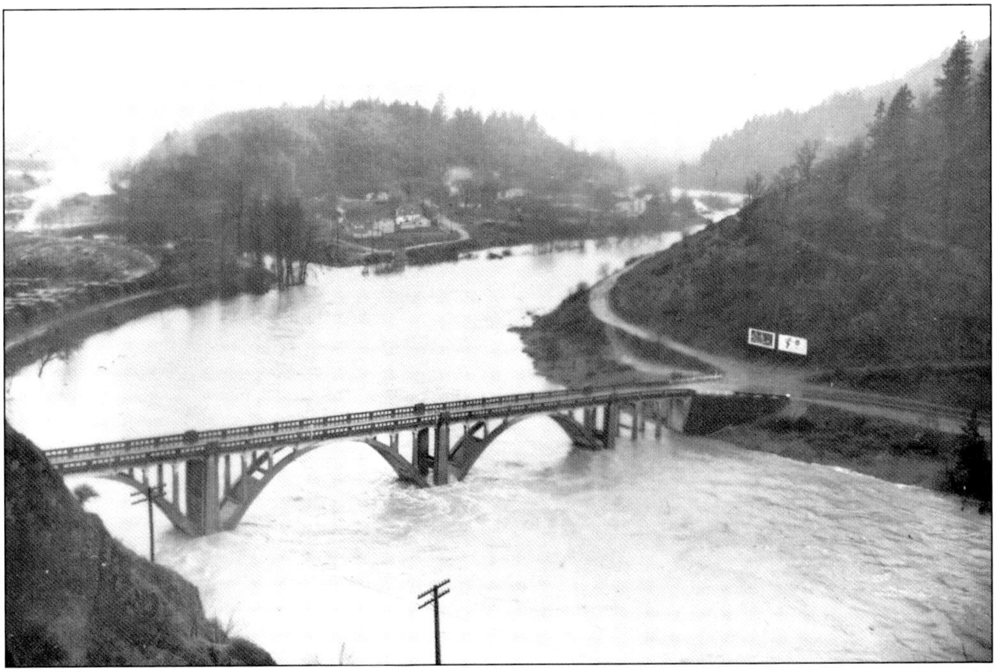

After this 1955 photograph was taken, the water continued to rise and washed out the cold deck of logs seen on the left bank above. The raging torrent hurled big logs against the nearly submerged bridge with devastating force, yet the bridge held, a testament to the engineering prowess of designer Conde McCullough. Ninety years after it was built, as graceful and beautiful as ever, the famous bridge holds fast.

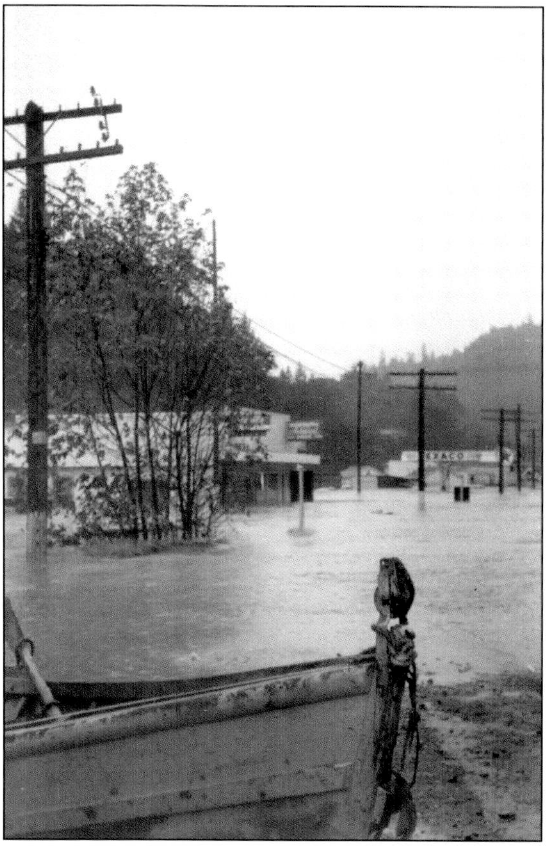

Highway 99 was deep in water in December 1955 when this photograph was taken on South Main Street across the creek from Umpqua Plywood. The only traffic through the water Thursday morning was by boat and lumber carrier. Interstate 5 was not completed at this time, so this effectively shut off all north-south traffic. Residents in the area vacated their homes at midnight Wednesday when the water started rising swiftly.

The Myrtle Creek Hospital and Clinic on Division Street opened on May 10, 1954. The hospital cost $100,000 to build, and Myrtlewood was used in the main reception area. After serving the community well for nearly 40 years, the hospital closed it doors around 1990.

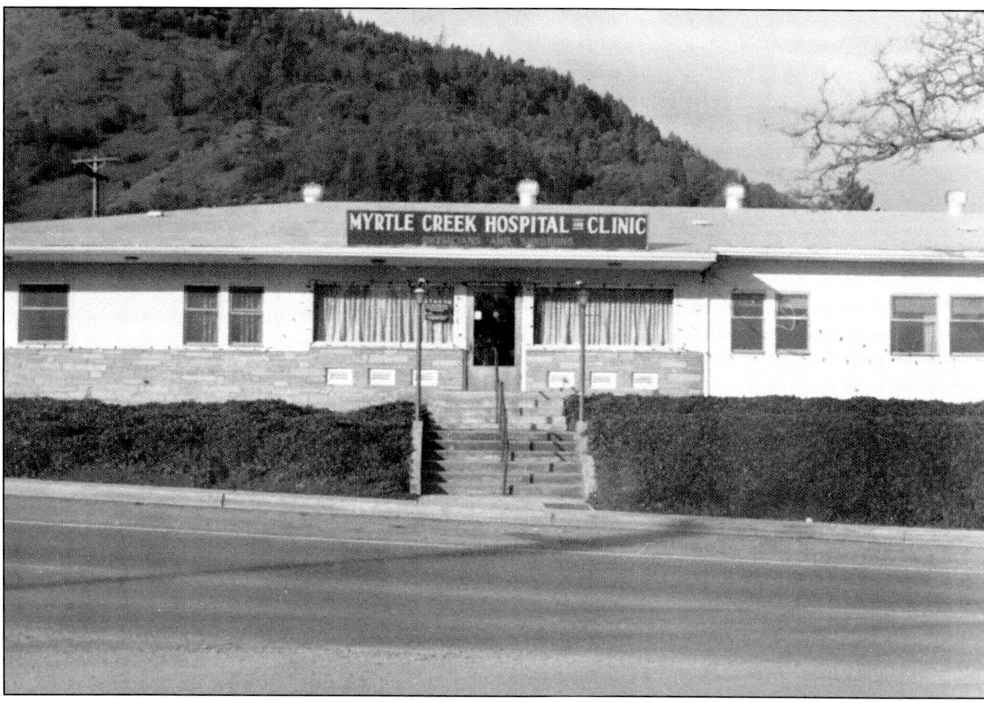

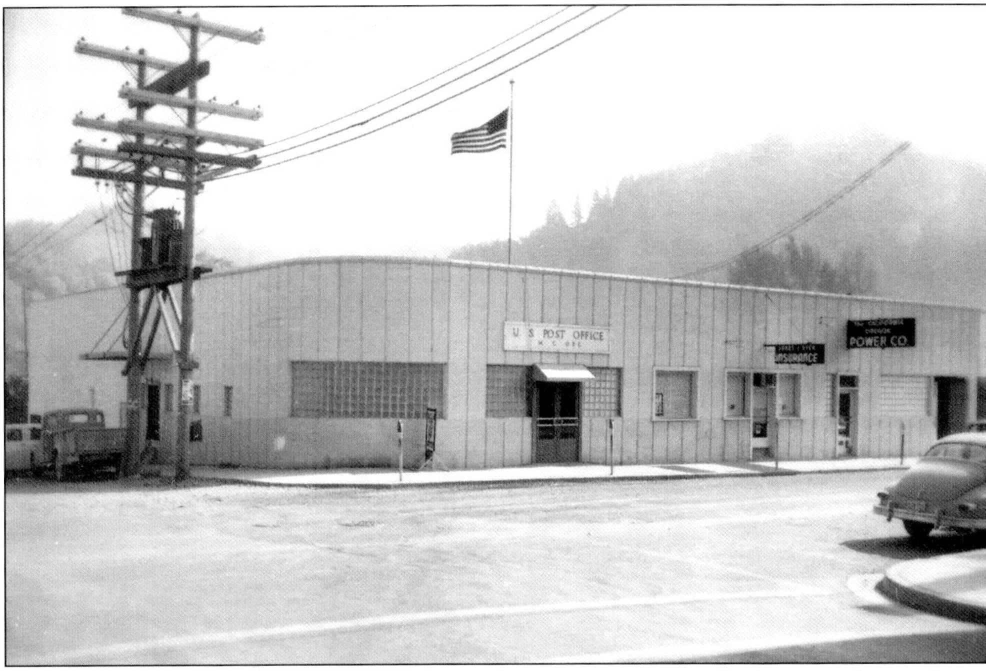

As previously mentioned, the Art Moderne style was prominent in the business district during the second quarter of the 20th century. Art Moderne is a term that describes advanced design ideas from the 1930s and 1940s. Many of these buildings have since been updated and wear a false facade that hides their original features. The above 1952 photograph of the US post office on First Avenue is a good example of the style. This building boasts sleek lines, curved walls, and horizontal bands of glass-block windows. This streamlined design continues in the photograph below of Jones Insurance and the California-Oregon Power Company. The empty space between the buildings at the end of the curved wall is the future home of the soon-to-be-built Rio Theater, a 1950s icon. (Both, courtesy City of Myrtle Creek.)

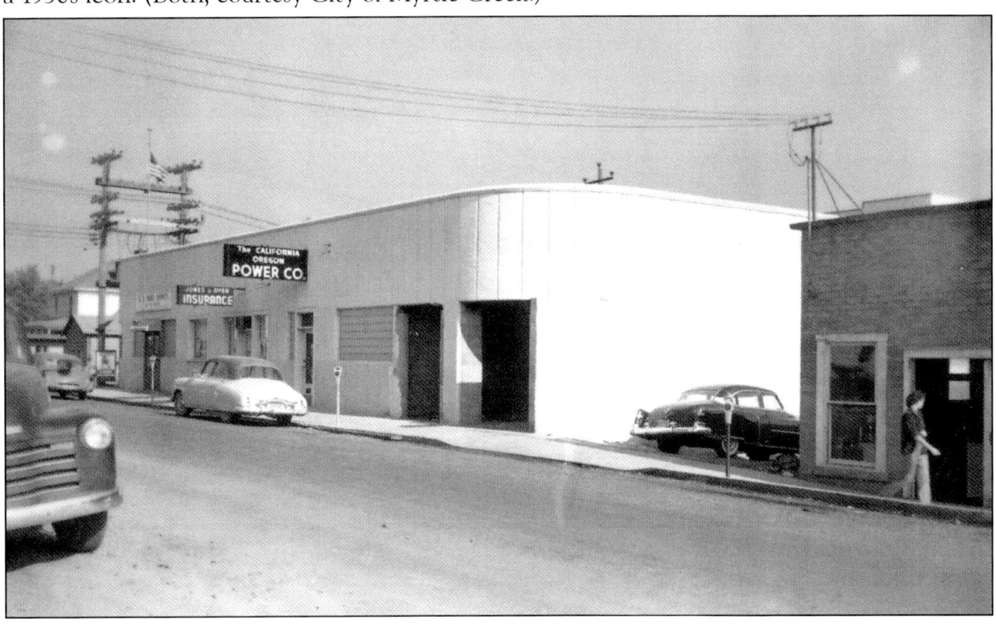

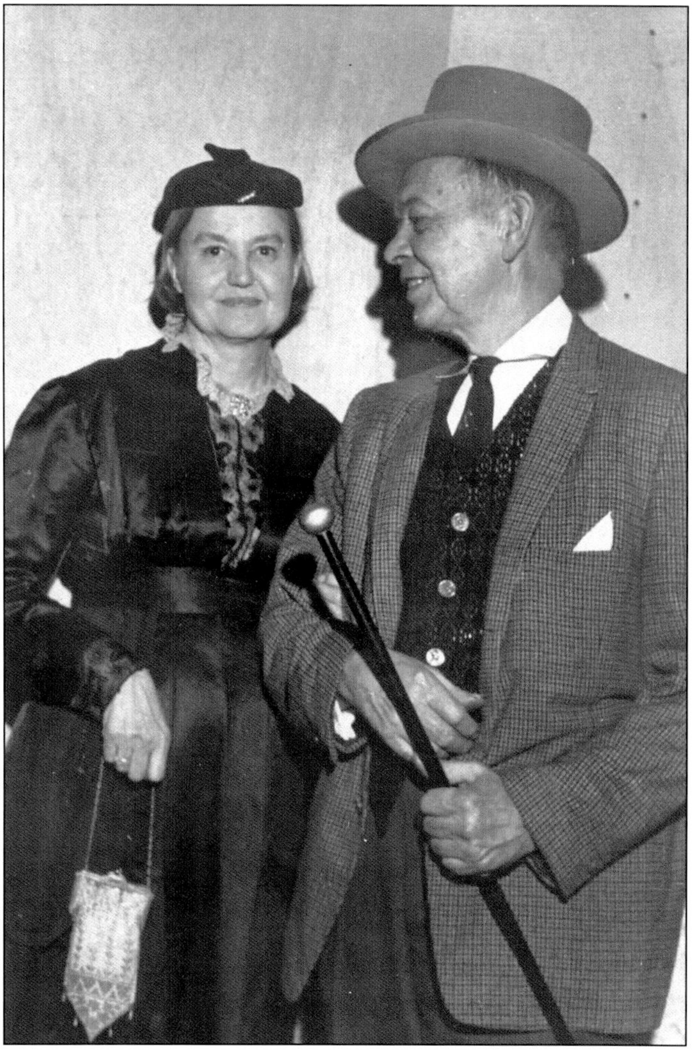

Hal and Portia Schiltz (left) were mainstays in the community for over a quarter of a century. Together they owned, edited, and published the local newspaper, the *Myrtle Creek Mail*, from 1948 to 1974. Supporters of a wide range of civic causes, they were particularly involved with library improvements. Based on the period clothing worn in the photograph, it was probably taken at one of the numerous civic events they attended. The Schiltz building on Second Avenue (below) was where the *Mail*, the *Umpqua Trapper*, and other publications were printed. Hal and Portia established a trust, transferring ownership of their building to the City of Myrtle Creek for public use. In the 1970s, the building was converted into a community center. (Left, Courtesy of Douglas County Museum; below, City of Myrtle Creek.)

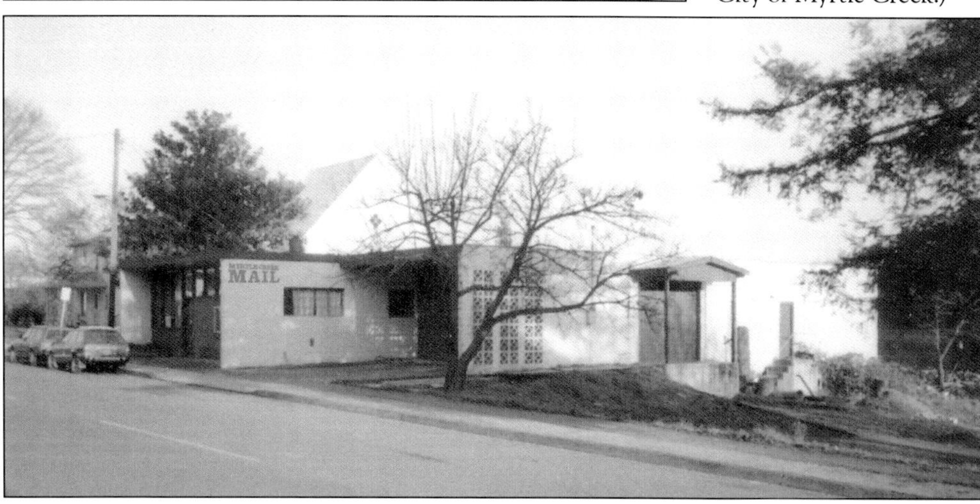

Modernizing of the municipal buildings on Pleasant Street began with the construction of a new police department around 1950 and continued in phases for about 10 years. Shown here sitting next to the ancient city hall is the main entrance of the newly built police department with a small office and even smaller booking room. The row of windows on the left are for—you guessed it—the jail cells. (Courtesy of City of Myrtle Creek.)

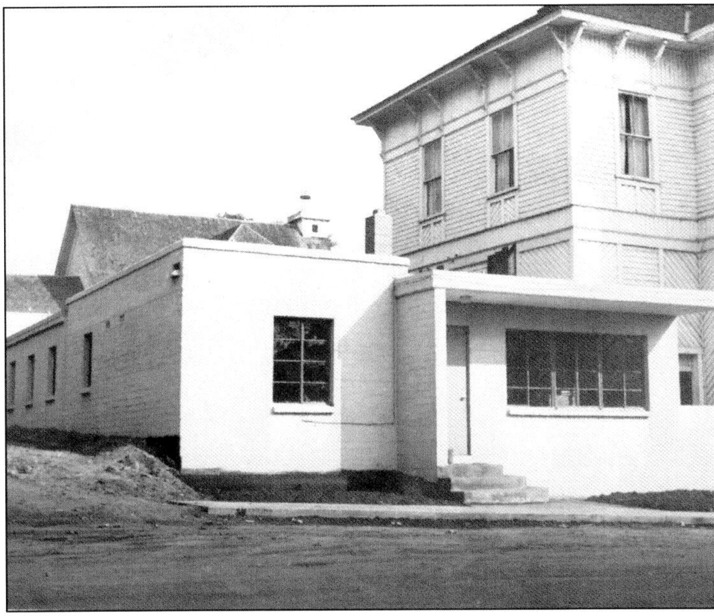

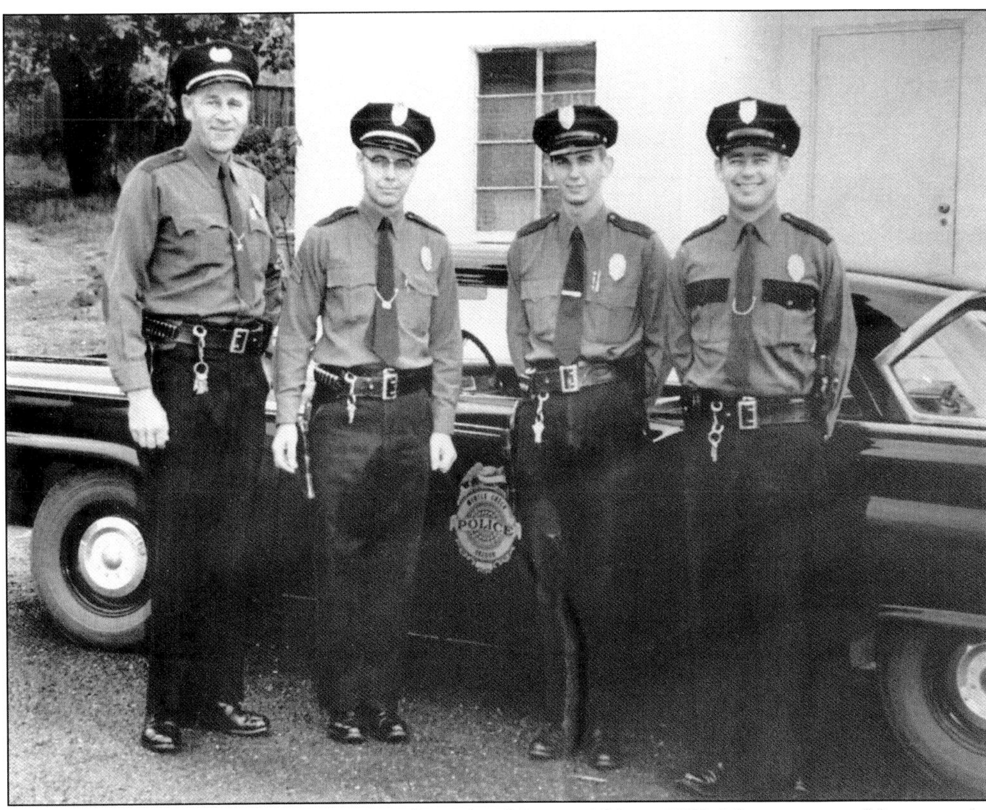

Jim Pringle (left) was Myrtle Creek police chief from 1956 to 1979. He is pictured here in the 1960s with, from left to right, Sgt. Russ Klumph, Jimmy Phillips, and Lloyd Freeman. (Courtesy of Myrtle Creek Police Department.)

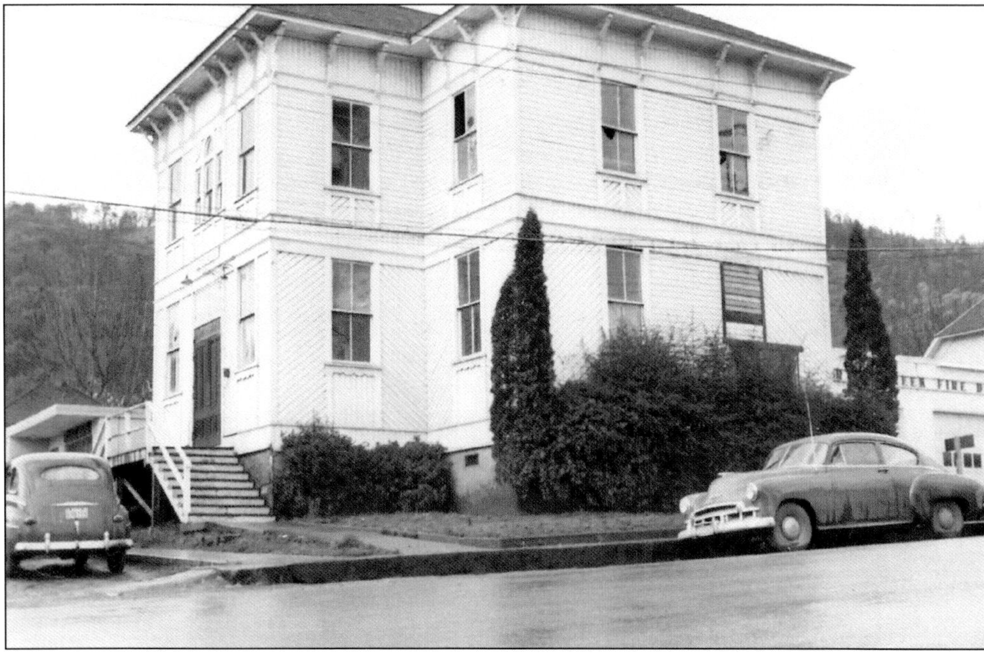

Missing its bell tower and showing its age in 1952, this is the same school building shown in chapter two that was once an appealing example of 19th century architecture. For 50 years, this building served as Myrtle Creek City Hall, but times and needs changed and plans were drawn for a new, modern city hall. (Courtesy of City of Myrtle Creek.)

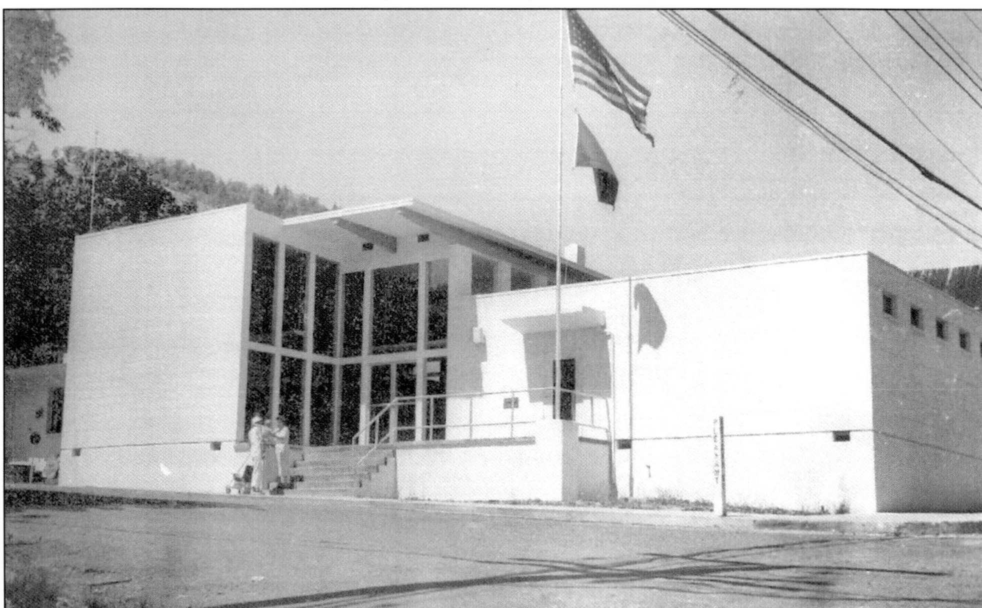

Completed in 1959, the new city hall is a good example of modern, mid-century architecture. Things have certainly changed since 1865, when John Hall first platted the town. Myrtle Creek incorporated as a city in 1883. When the last $2.94 in the treasury was spent in 1901, the city was broke and was forced to un-incorporate until it was solvent again 10 years later. (Courtesy City of Myrtle Creek.)

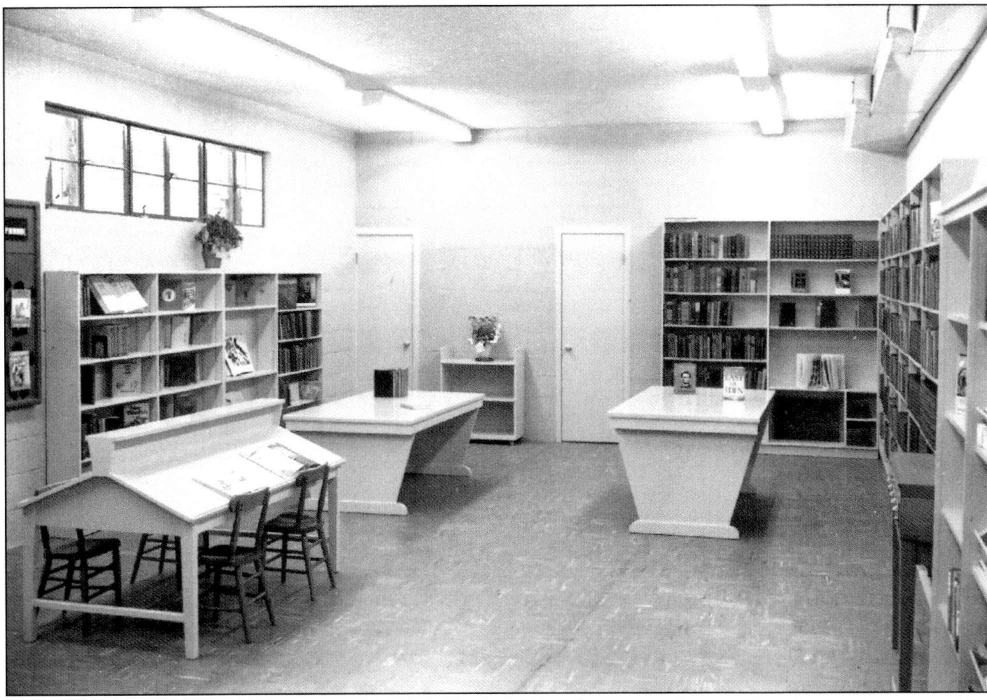

With a Second Avenue entrance, the city's first public library was located in the back room of the new police department building (see page 123). The photograph above shows the entire library in 1952. When the new city hall was built in 1959, the library moved into the east wing and was able to expand. Later, a new, modern library building (below) was built on Division Street as a branch of the Douglas County Library system. Then the east wing of city hall became the council chambers it is today. (Both courtesy City of Myrtle Creek.)

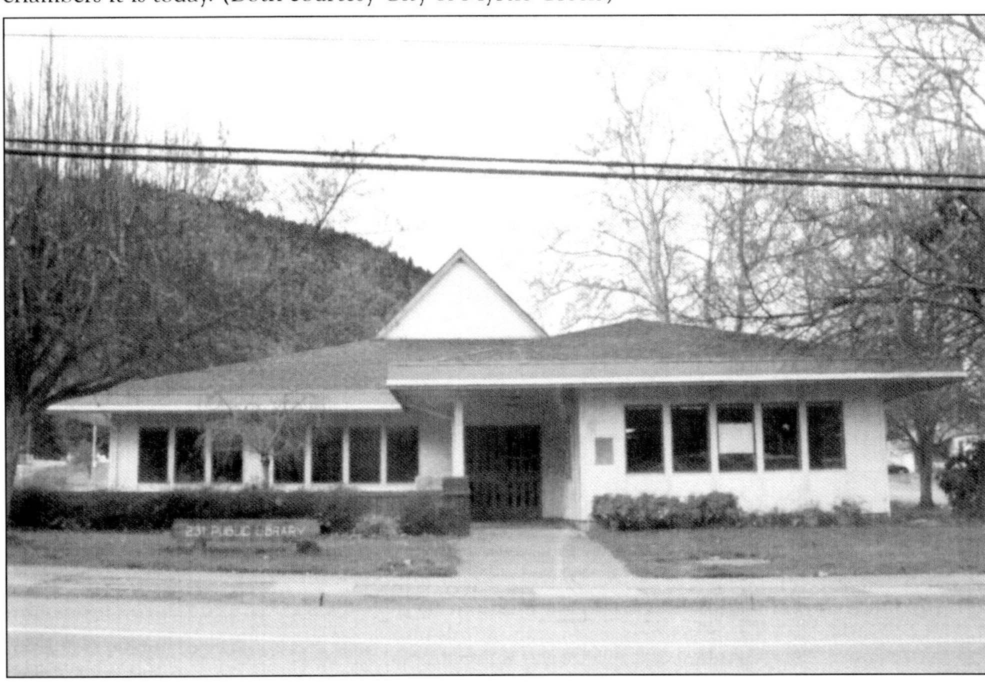

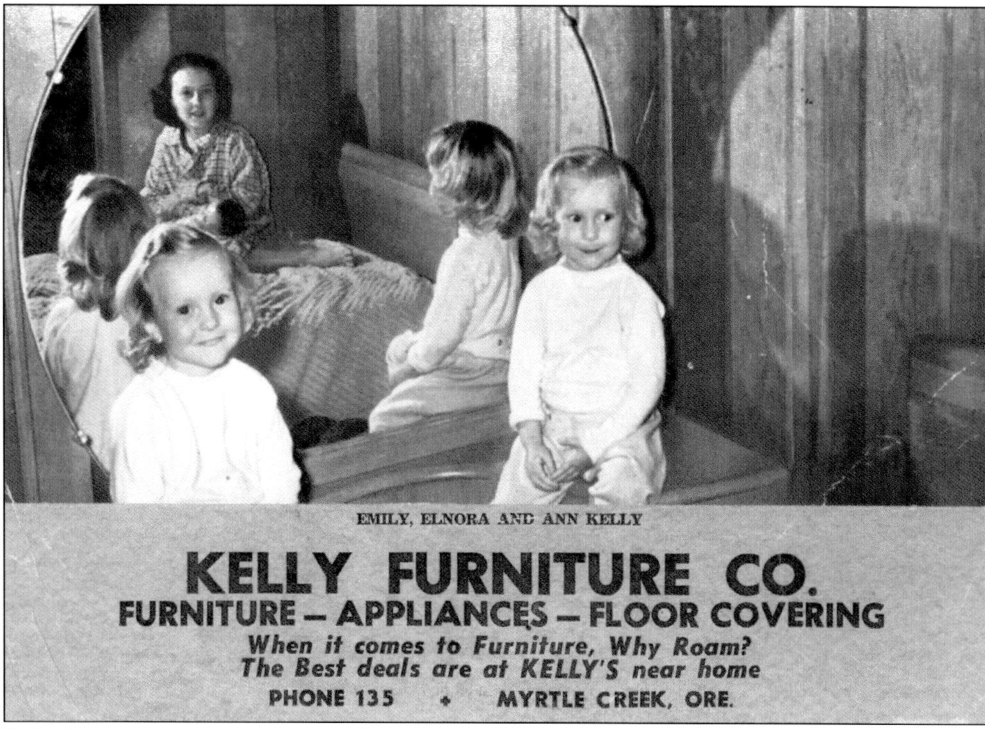

Kelly Furniture was a landmark for many years in downtown Myrtle Creek. This calendar advertisement shows the Kelly family sitting on the waterfall-style furniture that was popular in many homes from the 1930s to 1950s. Store owner Bob Kelly, the Lions Club, and many others were instrumental in helping raise funds to build the municipal pool shown below. (Courtesy of City of Myrtle Creek.)

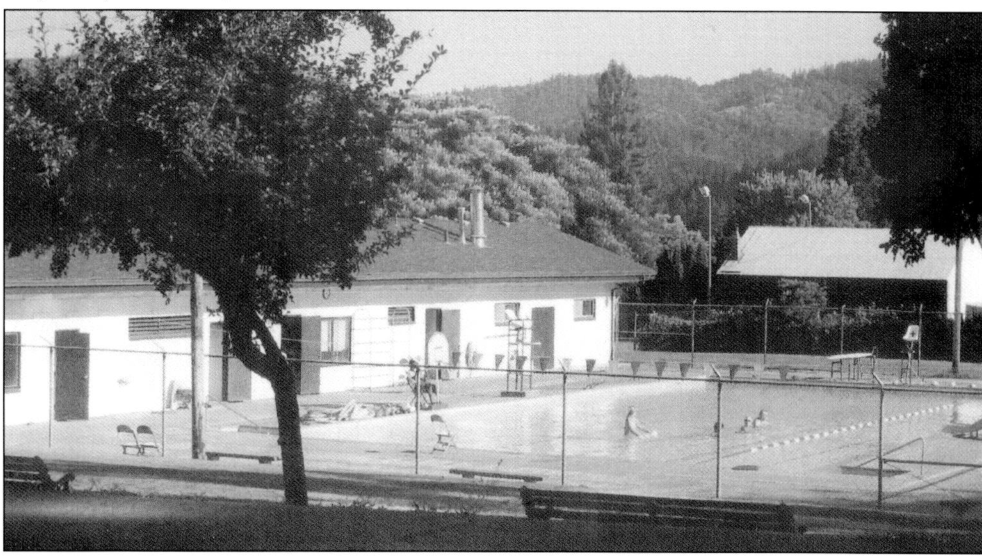

Bob Kelly was but one of many local citizens and civic groups prompted to build a municipal pool after a young girl drowned in the river. Living in the Seven Oaks house directly above the pool, Bob became the butt of much good-natured teasing when his car rolled down the hill one afternoon and plunged into the new pool. (Courtesy of Florence Chaney.)

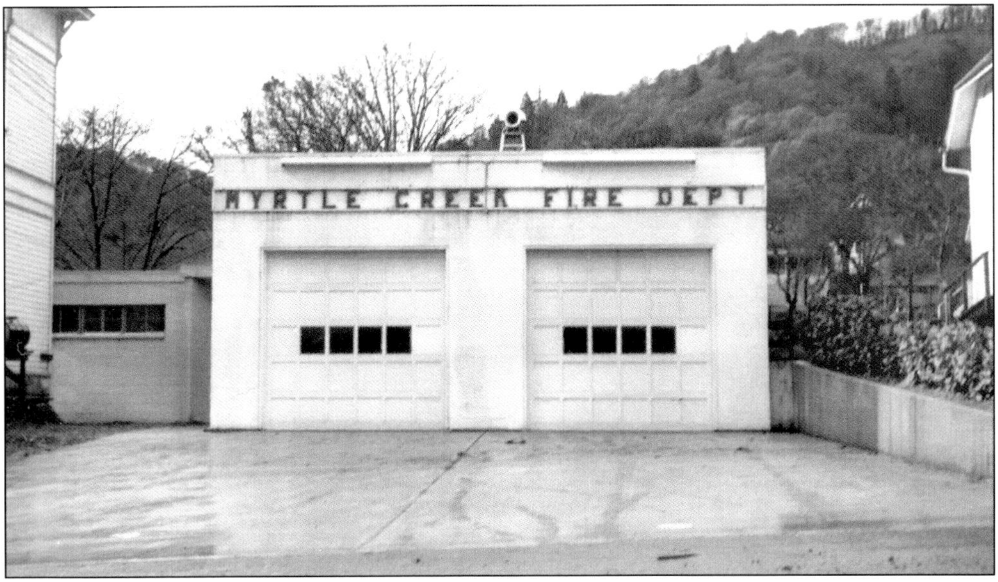

This humble hall was once the home of the Myrtle Creek Volunteer Fire Department. This wholly volunteer organization has served the city for over a century with a level of enthusiasm, public support, and firefighting expertise seldom attained by small town fire departments. They can be very proud, and rightfully so, of their selfless contribution to the community. (Courtesy of City of Myrtle Creek.)

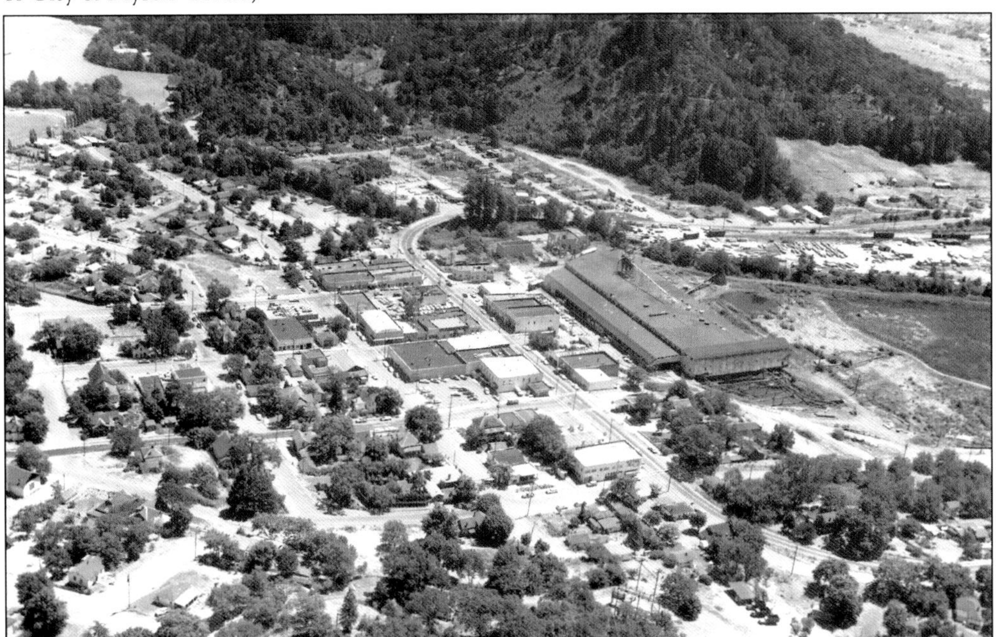

A victim of changing times, the mill phased out operations from the late 1950s through the 1960s. After a fire destroyed the remnants in 1970, the site was donated to the city for a park. No longer a bustling mill town, Myrtle Creek has its roots planted in timber and farming. It is our wish that this pictorial retrospective of our community's first 100 years be valued by generations to come. (Courtesy of City of Myrtle Creek.)